PHYSICAL PRINCIPLES OF EXPLORATION METHODS

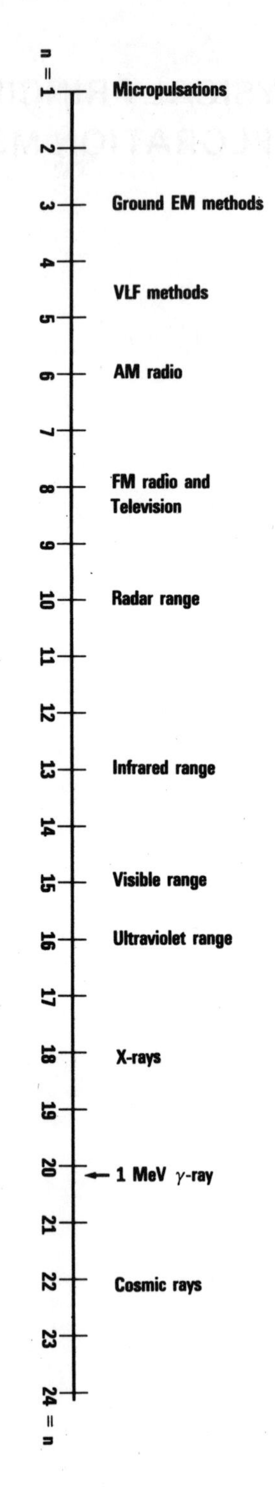

ELECTROMAGNETIC SPECTRUM
Frequency (in 10^n Hz)
n = 1
2
3
4
5
6
7
8
9
10
11
12
13
14
15
16
17
18
19
20
21
22
23
24 = n
Micropulsations
Ground EM methods
VLF methods
AM radio
FM radio and Television
Radar range
Infrared range
Visible range
Ultraviolet range
X-rays
1 MeV γ-ray
Cosmic rays

PHYSICAL PRINCIPLES OF EXPLORATION METHODS

An Introductory Text for Geology and Geophysics Students

A. E. Beck
Dept. of Geophysics, University of Western Ontario

A HALSTED PRESS BOOK

JOHN WILEY & SONS
New York

First published in Great Britain 1981 by
The Macmillan Press Ltd

Published in the USA by
Halsted Press, a division of
John Wiley & Sons, Inc
New York

LCCCN 81–80411
ISBN 0 470–27124–8 (cloth)
0 470–27128–0 (paper)

Printed in Hong Kong

To my wife Julia and the late J. C. Jaeger who between them have made me what I am and therefore, for better or worse, bear some responsibility for this book.

Contents

Foreword

This book is the outcome of a number of years of lecturing in Pure and Applied Geophysics to third and fourth year students drawn from a wide variety of disciplines but principally Geology, Physics and Geophysics. The difficulty in teaching a course to a class of such mixed backgrounds is the usual one of trying to avoid boring all of the people all of the time; I have long since given up trying to avoid boring some of the people some of the time.

The principal purpose of the course is not to produce a good geophysical interpreter but to produce students who understand enough of the basic physical principles involved in any given geophysical method that they will appreciate its limitations as well as its strengths. Once these important aspects are understood it becomes easier to select a particular method, or methods, in any given geological situation.

The book is aimed principally towards the geologically inclined student although experiences show that it is not harmful to the physicist if principles are stressed once again. Thus, where a development can be made simply as well as rigorously, this is done; however, if a rigorous development runs the risk of involving the student in too much boring detail, non-rigorous short cuts are taken without, I hope, losing sight of the basic principles involved. This might offend the purist but I make no apology for that since rigorous development can be left till later for those who are sufficiently determined that that is the direction they wish to go.

For similar reasons, and because of time limitations in the lecture room and space limitations in the book, I have attempted to convey some information by osmosis. For example, in the discussion of electromagnetic methods brief reference is made early on to electric currents being stimulated in conductors by a time varying magnetic field; the physicist would immediately appreciate the principle and think no more about it before moving on. Other brief references to eddy currents are made throughout the chapter, and indeed in other chapters, in slightly different contexts, and the non-physics student will, I hope, arrive at the end of the chapter accepting the principle even if he doesn´t fully understand how it comes about; from there it is but a small step to appreciate some of the problems involved in the electro-magnetic method.

Too often, a student is lead to believe that there is only one way to skin a cat; therefore in some places an alternative development is indicated. For example, in the chapters on Gravity and Magnetic Methods simple equations are developed using either the force field approach or the potential field approach, so that the student may come to realize that in problem solving of any sort there is a choice of technique, the particular selection depending on the problem and, often enough, on personal preference based upon familiarity.

In places the development may be heavy going for students without a first year calculus course under their belt; wherever possible physical principles and mathematical representation are illustrated by simple diagrams.

Although the emphasis of this book is on the physical principles underlying a particular method used in the field, techniques of interpretation are not entirely ignored since they too depend on understanding of the method used. However, there is no detailed examination of interpretation techniques; generally, only an outline of the reasoning is given.

Suggestions for further reading are made at the end of each chapter; they are not all embracing. Some are chosen to lead the reader to a classic paper while others lead to a recent publication with numerous references to earlier work; some lead the reader to a journal or book which may contain a number of interesting and relevant papers and, of course, some are given for the standard reason of allowing the reader to follow up statements made in the text.

One other tricky decision to make relates to questions of units. In the end I have decided to stay with the traditional c.g.s. since this is the system used in the great majority of literature to which students will refer; however, at strategic locations throughout the text, the recommended equivalent SI units are given in brackets.

Finally, I would not be unhappy if by the end of the book the student realizes that there is nothing magical or mysterious about geophysical exploration; that the methods are based upon firm, relatively simple, principles; that success depends on intelligent use of the methods available; and above all that good interpretation requires a careful marrying of the geophysical data collected to the known geology of the area.

Many people have read and commented on the text at various stages; they are too numerous to list all by name but I would particularly like to thank R.A. Facer, L. Mansinha, H.C. Palmer and D.M. Shaw for their critical comments, D. Graves and F.W. Graves who prepared the diagrams, and Cathy Fodemesi who has patiently prepared the text in copy ready form.

1

Introduction

It is as well to place Geophysics in context with the related disciplines, in which I include Chemistry as well as Geology and Physics. To do this, simple minded definitions are used with all the dangers inherent in brief and simple statements which are not hedged in by innumerable qualifying phrases.

Geology is basically the science pertaining to the history and development of the earth as deduced from a study of rocks. Chemistry, on the other hand, is the science of the properties of substances, both elements and compounds, and the laws of their combination and action on one another. Physics is the science of the properties (other than chemical) of matter and energy. Geophysics, therefore, is the study of the properties of matter and energy of the earth and earth materials.

Obviously, there are areas of considerable overlap between all these disciplines but particularly between Geology, Geophysics and Physics. This leads to a natural division of Geophysics into two main parts (a) Pure Geophysics which is basically a study of the physics of the solid earth regarding it as a continuum, and (b) Applied Geophysics, or Exploration Geophysics, where we apply physical principles and methods in a search for economic deposits.

Many people might quarrel with the tacit assumption that Geophysics is a discipline in its own right; however, I doubt whether many people would dispute that the Geophysicist is a professional in his own right. A reverse argument might be applied to Geology. Not many people would disagree that Geology is a separate discipline; but it can probably be argued that there is no such animal as a Geologist. A person practicing in the discipline of Geology these days usually has to be a pretty good Chemist, or Physicist, or even Mathematician; after all, a field geologist identifies and maps a rock by many methods. By sight - using an electromagnetic method where the frequency happens to be in the visible range; by chemical analysis when trying to identify carbonate material by dropping hydrochloric acid on to a specimen; when the pace and compass method is used in field mapping the observer is not only making a physical measurement (length) but is also using fundamental physical properties of the earth (its magnetic and gravitational fields).

Having demonstrated that there is no such thing as a Geologist, but that there are such people as Geophysicists and Geochemists, we can proceed with a general discussion of geophysical methods.

Purpose of Geophysical Exploration

Geophysical methods are used to gather information on subsurface geological structure. This information then forms part of a data set used in such diverse areas as soil mechanics investigations, developing building codes, preparing plans for disaster amelioration, investigating the origin of the earth, and searching for economic deposits ranging from gravels to oil and gas. When used in the last context, geophysical exploration has a dual purpose. (1) To detect potentially economic deposits; and (2) to eliminate potentially barren prospects from consideration for further detailed work using geological, geochemical or other geophysical techniques.

Basis of Geophysical Exploration

All geophysical exploration methods depend fundamentally on the presence of bodies of contrasting physical (and occasionally physical-chemical) properties. In fact we can stretch a point and say that the fundamental limitation on any method is the lack of sufficient physical property contrast. All other limitations are really caused indirectly by this. For instance, instead of saying an instrument is not sensitive enough we can say that the body is not sufficiently in contrast with its surroundings to be detected by that particular instrument. Therefore, tabulating the limitations we have

Main - lack of sufficient physical property contrast between the bodies of interest and the surrounding formations being studied.

Subsidiary - (a) Instrumental lack of sensitivity.
(b) Decrease of response with increase of distance between detector and body of interest.
(c) The limitation implied by (b) leads to an uncertainty in interpretation for we frequently cannot tell whether the response is due to a small body near the surface or a large, more dispersed, body deep down i.e. ambiguities in interpretation.

Response Curve

A plot of position versus abnormal response of a detector to the physical property anomaly of a body of interest yields what we call a response curve. Interpretation of a response curve is made by a comparison with ideal theoretical curves, assuming we have some knowledge of the body`s shape and physical characteristics. This means that we really rely on experience and essentially use the method of analogy. i.e. "what has worked somewhere else may work here if the general conditions are similar". Inevitably, this approach does not yield a unique solution and geological knowledge is required for a proper interpretation which at best only sets limits on the number of reasonable solutions. If the interpretations give a worthwhile indication this might mean that further work is required in the area including the possibility of drilling a prospect.

Factors Affecting a Decision on Exploration Work

A decision on whether an area should be further investigated depends upon many factors, most of them interdependent.

1. The size of the area to be covered
2. Type of terrain involved
3. Type of method that can be used
4. Cost of running a survey
5. Cost of production of the products being sought, both now and in the future
6. Market value of the product being sought, both now and in the future
7. The political stability of the area in which the deposit is located
8. Whether or not other ore, or oil, has been found in the area (there is some truth in the old mining proverb that the best place to look for ore deposits is in an area where some have already been found)
9. Availability of funds
10. Even the ability of the project coordinator to make a case has some influence in the decision making process and here people can have off days as well as on days.

No doubt there are a number of other factors which could be mentioned but the important point to note is that a decision on whether or not to proceed with a geophysical survey depends not only on the scientific merit and reliability of the method under consideration, but also upon many less tangible factors.

Required Physical Property Contrast for Various Methods

Since all exploration methods depend on physical property contrasts we must consider which physical properties are associated with a particular method. Sometimes the relevant physical property is only indirectly involved with the quantities we actually measure; sometimes more than one physical property is involved and we can only speak about physical property ratios. Table 1 summarizes the situation with regard to all the commonly used methods. Only the commonly used phenomena are indicated in

Table 1.1 Basis of geophysical exploration methods.

Method	Depends on changes in (Basic Physical Properties)	Makes use of (Phenomena Involved)	Direct or Indirect*
Electrical	electrical conductivity (σ)	apparent conductivity	D
Spontaneous polarization (SP)	oxidation potential (Eh) hydrogen ion concentration (pH) electrical conductivity (σ)	electrochemical potentials	I
Induced polarization (IP)	electrochemical properties of electronically conducting particles in rock pores: ion concentration in pore fluids	polarization voltages	I
Electromagnetic	electrical conductivity (σ) magnetic permeability (μ)	phase and intensity relationships of alternating magnetic field; alternating electric field	D
Gravity	density (ρ)	spatial variations in gravitational acceleration (force)	D
Magnetic	magnetic susceptibility (k) (or permeability)	spatial variations in stationary magnetic field (force)	D
Seismic	elastic constants, density	seismic velocities of compressional waves	I
Radioactive	abundances of radionuclides	gamma radiation activity	D
Thermal	thermal conductivity	heat flow	I

*Method is classified as indirect (I) if little or no attempt is made to obtain the contrast (or absolute value) for the physical property involved.

Table 1. For instance, there have been a number of attempts to use shear wave velocities and amplitude ratios in the seismic method. However, although increasing use is being made of amplitude ratio studies, particularly in crustal seismology, the vast majority of interpretations are still made basically on travel time data for compressional wave velocities.

The methods can be grouped in different classes. For instance, they could be described as either active or passive. In the active method we rely on stimulating a response by artificially energizing the ground and sub-surface; for example, in the electromagnetic methods we use EM waves, usually in the frequency range 50 to 5000 Hz. On the other hand, in the passive methods we simply rely on detecting variations in naturally produced fields by introducing a suitable exploring instrument (e.g. gravimeter or magnetometer).

The methods could also be grouped according to whether they are static or dynamic. In static methods we measure the spatial variation of a stationary (non time-varying relative to the duration of the survey) field, an example being the use of the magnetometer in the magnetic method. In dynamic methods we measure the spatial variation of the time varying field; thus the electromagnetic method could be called an active-dynamic method. On the other hand AFMAG, which relies on naturally generated EM noise pulses (sferics) could be called passive-dynamic.

Signal, Message and Noise

Finally, in this introductory chapter we must mention the distinction between message and noise in the signal.

Signal = message + noise.

Three types of noise can be distinguished, instrumental, operator and geological. With modern equipment, instrumental noise is usually insignificant compared with the other two types. Operator noise, that is the error introduced into the measurement by the operator of the instrument, can usually be kept small by careful field practices; however, some types of field equipment are particularly prone to operator error - e.g. misalignment of the field coils in the horizontal loop electromagnetic methods.

Geological noise may be broadly defined as the contribution, to the total response, of geological bodies or formations which are not of interest to the operator.

In this context it should be realized that, assuming negligible instrumental and operator noise, all information in the signal has a meaning. The difficulty is in understanding what the signal has to say. Therefore, unwanted portions of the signal (noise) are often removed to leave a particular portion which is easily understood (message). However, it is most important to recognize that one man´s noise might be another man´s message. For instance, the commonly encountered Rayleigh wave is usually regarded as noise ("ground roll") by the exploration seismologist but may contain an essential message for the crustal seismologist, especially when the waves are in the longer wave length range.

SUGGESTIONS FOR FURTHER READING

These two books cover most of the methods discussed in this book but give a much extended treatment of some theoretical aspects and interpretation of field data.

Telford, W.M., Geldart, L.P., Sheriff, R.E. and Keys, D.A., 1976. Applied Geophysics. C.U.P. Cambridge.

Dobrin, M.B., 1976. Introduction to Geophysical Prospecting, Third Edition, McGraw-Hill, New York.

The following two journals should be available from any self respecting library supporting a solid earth geophysics group.

Geophysics. The journal of the Society of Exploration Geophysicists. Tulsa, Oklahoma.

Geophysical Prospecting. The journal of the European Association of Exploration Geophysicists. The Hague, The Netherlands.

2

Electrical methods

BASIC EQUATIONS

Experimentally we know that if we apply a voltage V across the ends of a body of constant cross section, the current (I) is proportional to the applied voltage (V), so that

$$V = RI \qquad \text{or} \qquad R = V/I \tag{2.1}$$

where R is the constant of the proportionality and is called the resistance of the body. We also know that for a given material the resistance is proportional to its length and inversely proportional to the cross sectional area so that

$$R = \rho L/A \qquad \text{or} \qquad \rho = R.A/L \tag{2.2}$$

where ρ is called the resistivity of the material and is the resistance of a cube with a side of unit length. The units of resistivity are usually given as ohm - meters (Ω-m) or, more pedantically, $\Omega\text{-m}^2/\text{m}$; this latter system follows from the dimensions of the right hand side of equation 2.2. The formal definition of resistivity is that it is the ratio of the voltage gradient to the current density over a small thin surface element of the medium.

The electrical conductivity is the inverse of the resistivity.

ELECTRICAL PROPERTIES OF ROCKS

The resistivity of naturally occurring rocks and minerals covers several orders of magnitude; if we include refined elements such as copper the range is further extended by a few orders of magnitude as shown in Table 2.1.

The figures in Table 2.1 are, of course, for the most part only approximate. Many factors can influence the electrical conductivities of materials in bulk. Such factors include porosity, degree of saturation of porous materials, nature of the pore fluids, types of minerals making up the rock matrix, nature and size of the grains making up the matrix, whether the material is well consolidated (this having an influence on the intergranular pressure and therefore contact resistance), the bulk porosity of large volumes (by which we mean that even in tightly welded formations, zones of crushing, fractures and faults, which may or may not be filled with electrolytes, will have an influence on the gross resistivity), and structure (both on the micro and macro scale since this may lead to significant anisotropy and inhomogeneity).

TABLE 2.1

Typical Resistivities for Some Common Materials

Material	Resistivity (Ω-m)
Copper	10^{-8}
Graphite	10^{-4} to 10^{-5}
Pyrrhotite	10^{-2} to 10^{-3}
Ore bodies (e.g. pyrite, magnetite, galena)	1 to 10^4
Wet limestone	10^2 to 10^3
Quartzite	10^{12}
Clays	1 to 100
Moraine	10 to 10^4

Because of the great range of electrical resistivities (conductivities) which can occur in nature, and therefore the large influence that small "impurities" can have on the results for gross conductivity, the electrical sounding method can only be used in a quantitative way in the simplest of geological situations. One of the simplest geological situations is that of a number of layers with boundaries between the layers being parallel to the surface over which electrical measurements are being made; if there is a lateral inhomogeneity, and this includes the case where an interface dips at more than 10° to 15°, it is almost impossible to interpret the data quantitatively. However, it may be possible to analyze the data qualitatively, as will be discussed later.

BASIC EQUATIONS

From equations 2.1 and 2.2 we have $I = VA/\rho L$

Now consider a point source in a plane which is the bound between a perfect-insulator and a semi-infinite, isotropic, homogeneous conductor of resistivity ρ, Fig. 2.1.

Resistance of the hemispherical shell of radius x is, from 2.2.

$$R = \frac{\rho dx}{2\pi x^2}$$

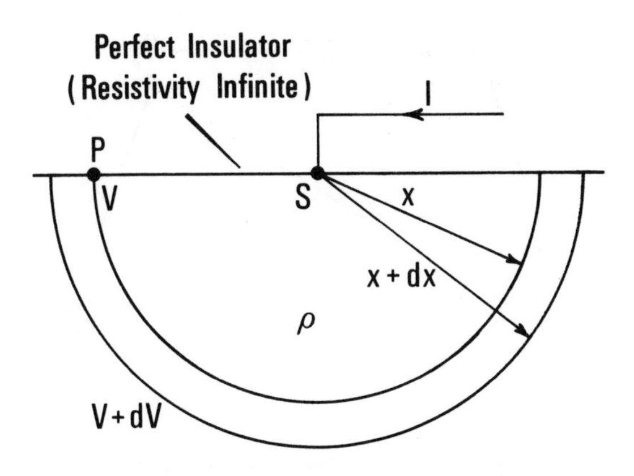

Figure 2.1. Section of hemispherical equipotential surfaces from point current source on a semi-infinite, homogeneous, isotropic body.

Invoking Ampere`s rule and Ohm`s law, the current flowing across the shell is therefore

$$I = \frac{V - (V + dV)}{\rho dx/2\pi x^2} = \frac{2\pi x^2}{\rho}\frac{dV}{dx}$$

or

$$dV = -\frac{\rho I dx}{2\pi x^2}$$

But from the definition of potential at a point (namely that the electrical potential at a point is the work done against the electric field in bringing a unit positive charge from infinity to that point) the potential at P is given by

$$V_p = \int_\infty^x dV = -\frac{\rho I}{2\pi}\int_\infty^x \frac{dx}{x^2} = \frac{\rho I}{2\pi}\left[\frac{1}{x}\right]_\infty^x$$

or

$$V_p = \rho I/2\pi x \qquad (2.3)$$

This is then the potential at the point P due to a point current source at S.

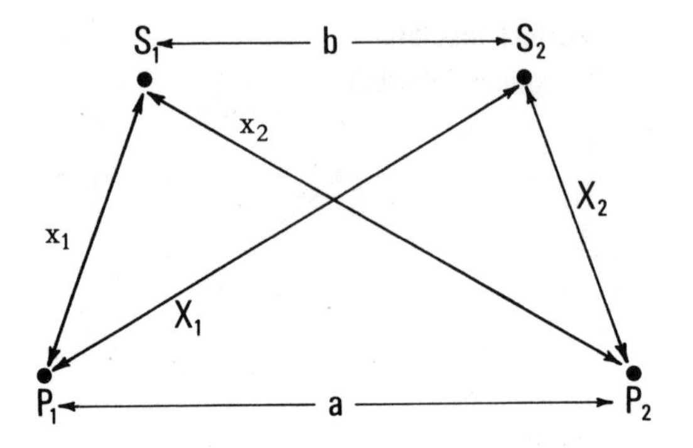

Figure 2.2. Plan view of generalized four electrode array.

Now suppose there are two current sources S_1 and S_2 of strengths I_1 and I_2 respectively. Consider the potentials at two other points P_1 and P_2 where the respective distances are as shown in the plan view of diagram of Figure 2.2.

The potential at P_1 due to S_1 is $\rho I_1/2\pi x_1$, and due to S_2 is $\rho I_2/2\pi X_1$.

The total potential at P_1 is therefore $V_{p1} = \frac{\rho}{2\pi}\left[\frac{I_1}{x_1} - \frac{I_2}{X_1}\right]$

Similarly the potential at P_2 is $V_{p2} = \frac{\rho}{2\pi}\left[\frac{I_1}{x_2} - \frac{I_2}{X_2}\right]$

The potential difference V between P_1 and P_2 is therefore given by

$$V_{p1} - V_{p2} = V = \frac{\rho}{2\pi}\left[I_1\left(\frac{1}{x_1} - \frac{1}{x_2}\right) + I_2\left(\frac{1}{X_1} - \frac{1}{X_2}\right)\right]$$

If S_1 and S_2 form a dipole system such as would be obtained by connecting S_1 and S_2 to the two terminals of a battery, then we can assume that all current entering the ground via S_1 eventually leaves via S_2, and $I_1 = -I_2 = I$ (say), so that

$$V = \rho I \frac{1}{2\pi}\left[\frac{1}{x_1} - \frac{1}{x_2} - \frac{1}{X_1} + \frac{1}{X_2}\right] \tag{2.4}$$

The term in square brackets can be regarded as the contribution of the geometry of the electrode system to the observed voltage; its inverse may be defined as the "array constant". The 2π term arises because we are dealing with a

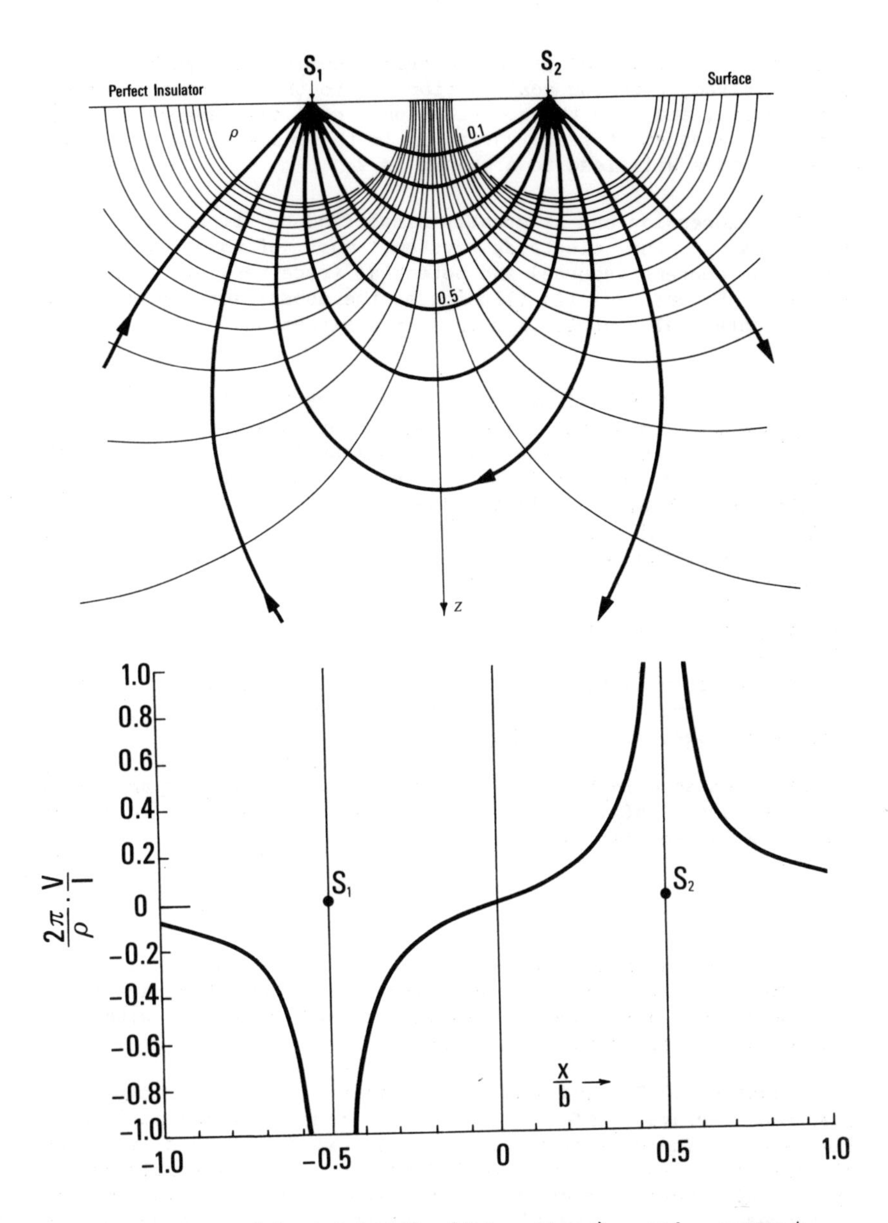

Figure 2.3. (A) Potential (thin lines) and current distribution in a vertical plane through a dipole source. There is a constant potential drop between successive equipotential lines. (B) Potential distribution on the surface due to dipole source. Adapted from van Nostrand and Cook (1966).

semi-infinite solid, and is the contribution of the geometry of solid involved, ρ is the contribution due to the geological nature of the solid, and I is the contribution from the energy source used. For a constant ρ any variation in the geometrical or energy source factors will lead to a change in V.

The expression can be greatly simplified by imposing simple geometries on the electrode system, most of which consist of having the current (source) and potential (receiver) electrodes in a line. Figure 2.3 (A) illustrates the subsurface current flow and equipotential distribution in such a situation.

In Figure 2.3(A) if two successive current contours are taken and projected into the third dimension by rotation about a line through S_1S_2 so that we have half a "doughnut", 10% of the current flow is in this doughnut. For a plan view the mirror image about the line through S_1S_2 should be added. Figure 2.3(B) shows the surface potential distribution along the line joining the dipole sources; note the relatively constant gradient of potential over the range $-0.15 < x/b < 0.15$.

IMPORTANT TYPES OF ARRAYS

Two types of arrays are in common use, the Wenner array, mainly in North America, and the Schlumberger array, to be discussed later, mainly in Europe.

For instance in the Wenner array all the electrodes are in a line, Figure 2.4(A), with $x_1 = x_2 = a$, $b = 3a$ and therefore $X_1 = X_2 = 2a$, so that $V = \rho I/2\pi a$, or

$$\rho = 2\pi a \; (V/I) \tag{2.5}$$

The term in brackets has the form of equation 2.1 and therefore the dimensions of "resistance"; in fact there are field instruments which measure the ratio of V/I directly with the readout being called the "resistance".

Equation (2.5), and similar equations for other types of electrode arrays, refer only to the ideal case illustrated in Figure 2.1. The real earth is not so kind as this and the electrical resistivity, or conductivity, of earth materials varies by several orders of magnitude, sometimes over relatively short distances. It is therefore almost impossible to obtain a true value of the resistivity of the subsurface.

However, the array methods can still be used since it is obvious that no matter how complex the variations in subsurface resistivity, if a current is passed into the ground, through two electrodes, a potential difference can be measured between two other electrodes. The ratio of the potential difference to the current, multiplied by the array constant, will give a quantity with the dimensions of resistivity; this quantity is called the apparent resistivity (ρ_a) of the subsurface and is a function of the geological resistivity structure within the range of significance and resolution of the technique being used. Variations in ρ_a therefore indicate changes in the subsurface conditions and in some relatively simple situations, e.g. horizontally stratified structures, the variations in ρ_a can be used to give quantitative information, such as the thickness and true resistivities of the horizontally layered units.

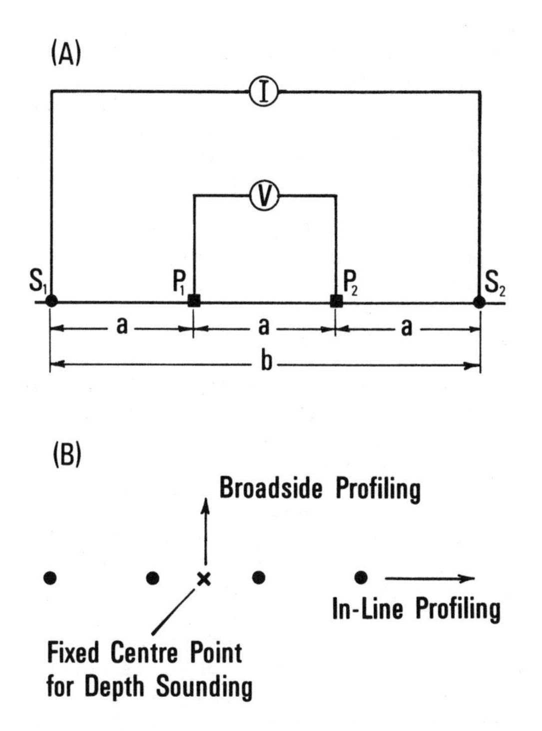

Figure 2.4. (A) Wenner array. (B) Three ways of using the array.

FIELD USE OF ARRAYS

Arrays can be used for either profiling or depth sounding, Figure 2.4(B), often referred to as electric "trenching" or "drilling" respectively.

Profiling (or "electric trenching")

In profiling, the four electrodes can be moved either on the line of the direction of movement (the in-line approach), or along a line which is perpendicular to the direction of movement (the broadside appraoch). In either case, it is easy to see how a change in the electrical resistivity could change the appearance of the apparent resistivity - position curve, Figure 2.5(A). In both cases the interpretation is nearly always qualitative. For instance, with a fixed spacing, it is clear that when the array is close to the surface boundary of the two materials most of the current flow will be in medium 2 to the left of the diagram, and when the array is far to the right, most of the current flow will be in medium 1. Thus a profile of apparent resistivity vs position would have the appearance given in Figure 2.5(A); with the change of the abscissa from position to electrode spacing this is similar in appearance to Figure 2.5(B). Nevertheless, if it is now assumed that the boundary is hidden under a thin layer of overburden, that is, the thickness is small compared with the electrode spacing, a qualitative interpretation can be useful.

In a few very simple cases, or when the problem is of sufficient interest to warrant model experiments, it may be possible to obtain quantitative results.

Depth Sounding (or "electric drilling")

When the array is used in the depth sounding mode it is usual to select a convenient fixed point which then becomes the centre of the array which is expanded about the centre, usually in equal increments of distance between adjacent electrodes. In this case it is easy to see qualitatively how the apparent resistivity - electrode spacing curve will appear, Figure 2.5(B). When the electrode spacing is small compared with the thickness of the first layer most of the current will flow in the first layer so that the influence of ρ_2 is relatively small; on the other hand, when the electrode spacing is large compared with the thickness of the first layer, most of the current will be in the second layer so that the influence of ρ_1 is relatively small. It is obvious that the change in the apparent resistivity - electrode spacing curve must occur smoothly to make the transition between ρ_1 and ρ_2. Although this is only a qualitative approach it is possible to obtain an approximate value of the depth to the discontinuity from the point of inflection of the curve. The actual position of the point of inflection will depend upon a number of factors such as the resistivity contrast and the thickness of the first layer but the depth estimate so obtained is a reasonable first guess.

Wenner Array. More quantitative methods have been developed for layered cases. The first widely used quantitative solutions were devised by Tagg in 1934 for the two layer (one discontinuity)

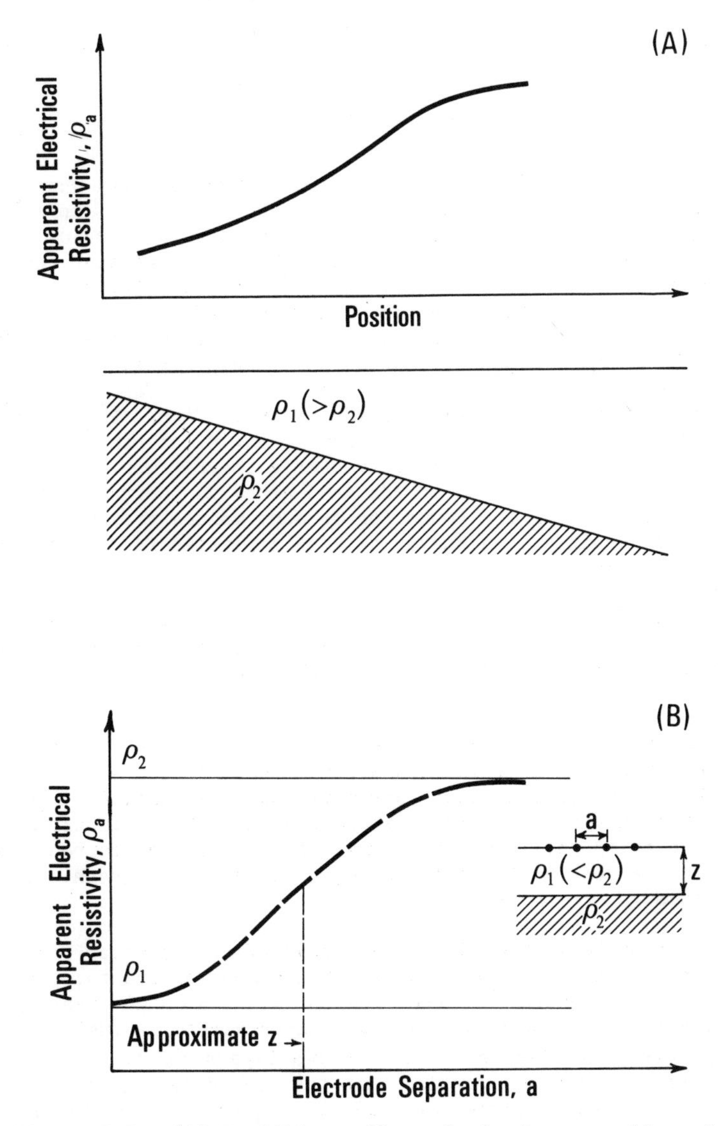

Figure 2.5. (A) Profiling. When electrode separation is large compared with depth to interface, the influence of ρ_2 is relatively large; when the electrode is small compared with depth to interface, the influence of ρ_2 is small. (B) Depth Sounding. For infinitesimally small electrode spacing the true ρ_1 is obtained while for infinitely large electrode spacing the true ρ_2 is obtained; but for spacing in between, an apparent resistivity ρ_a is obtained, where $\rho_1 < \rho_a < \rho_2$.

case. He related ρ_a to ρ_1 and ρ_2 via an equation of the form

$$\rho_a = \rho_1 . f(k, \frac{z}{a}) \qquad (2.6)$$

where $k = (\rho_2 - \rho_1)/(\rho_2 + \rho_1)$ and is called the resistivity contrast, or reflection factor since many of the theoretical approaches invoke the method of images; clearly k can have all values from +1 to -1, the positive value occurring when ρ_2 is greater than ρ_1. If more than one layer is involved the first resistivity contrast is usually labelled k_1, the second k_2 etc. z is the depth to the interface and a is the electrode separation. Using equation (2.6) Tagg calculated a number of curves which help fix z. Two sets of curves are used, one for $\rho_2 > \rho_1$ and another for $\rho_2 < \rho_1$. To illustrate the use of Tagg's curves refer to Figure 2.6; a value for ρ_1 is found by extrapolating back to a = 0 on the ρ_a vs a curve, then for convenient values of a the ratio ρ_1/ρ_a is found and tabulated. The values of z/a (and hence z) corresponding to each value of ρ_1/ρ_a are found for as many values of k between 0.1 and 1.0 as possible, and curves of k vs z plotted for each value of a that has been used.

In the ideal case it will be found that all lines cross in a particular point and the value of z at this point is the thickness of the top layer; the value of k at this point will give ρ_2 since ρ_1 is known.

Essentially, the above process solves the complicated equation 2.6 by a trial and error method, for once ρ_1 has been found by extrapolation, there are still two unknowns, ρ_2 (or k) and z, but only one equation. Therefore various combinations of z and k are tried until the solutions converge to the correct one. This can be done analytically using a computer, or graphically as done above.

Unfortunately, the ideal field conditions are never found and the curves intersect each other at a number of points; in this situation the centre of the region with a high density of intersections is selected as the approximate value of z. In appropriate cases Tagg's curves for two layers can be applied to the three layer case provided that the second layer, of thickness z_2, is sufficiently large compared with z_1 that the effect of the lowest layer can be ignored in the early part of the ρ_a vs a curve and that in the later part of the curve the value of a is sufficiently large with respect to z_1 that the contribution of the first layer to the apparent resistivity can be ignored, so that the second discontinuity can be found. The method works best when $\rho_1 > \rho_2 > \rho_3$.

However, in these circumstances it is better to use the

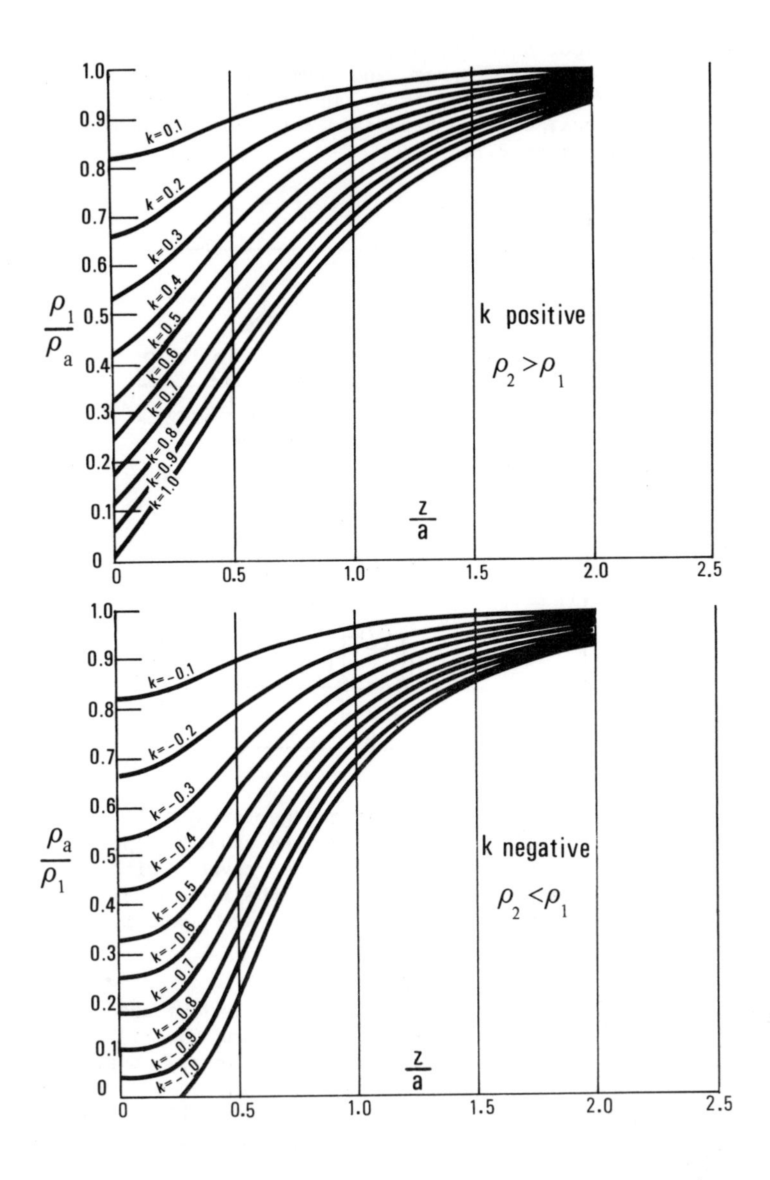

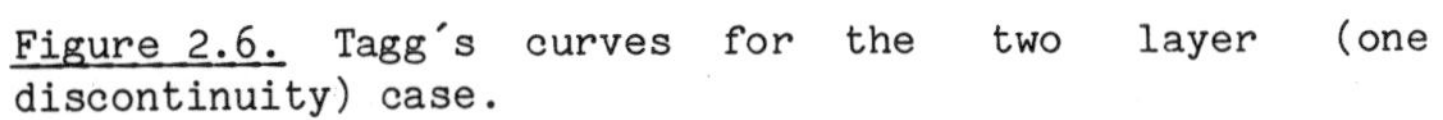

Figure 2.6. Tagg's curves for the two layer (one discontinuity) case.

Orellana and Mooney (1966) curves which are families of theoretical curves which have been developed for 2, 3 and 4 layer cases using selected layer resistivity and thickness ratios. Two groups of curves have been published, the first for the Schlumberger-electrode array and the second for the Wenner electrode array. The use of these curves, in an essentially curve fitting technique, will be discussed later using the Schlumberger array shown in Figure 2.7(B).

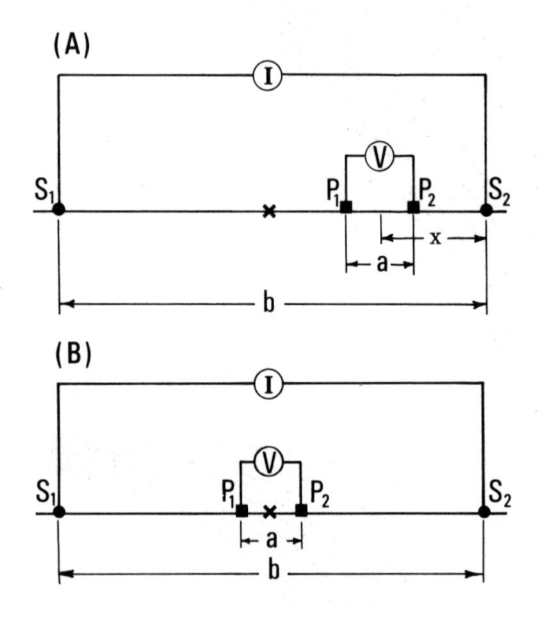

Figure 2.7. Schlumberger array. (A) For profiling S_1 and S_2 are usually kept fixed while P_1 and P_2 are moved, with constant separation, along the line. (B) For depth sounding P_1 and P_2 are usually

Schlumberger Array. The Schlumberger array may be used for both profiling and depth sounding. In profiling the field procedure can be similar to that used with the Wenner array, that is, all four electrodes are moved either in line or broadside with a constant, symmetrical array geometry; more usually S_1 and S_2 are kept fixed while P_1 and P_2 are moved along the line keeping a constant.

From equation 2.4 and the electrode spacings shown in Figure 2.7(A) it can readily be found that for the Schlumberger array the resistivity is given by

$$\rho = \frac{\pi}{2} \cdot \frac{V}{I} \cdot a\left[\frac{1}{4(b-x)^2-a^2} + \frac{1}{4x^2-a^2}\right]^{-1} \qquad (2.7)$$

In the depth sounding mode the potential electrodes are usually kept fixed in position and the current electrodes moved symmetrically with respect to the mid point of the symmetrical array, that is $x = b/2$ in equation 2.7. Often the value of a is small compared to b and the increments in b are in integral multiples of 2a. If this is the case and $b = (2n + 1)a$, where n is an integer, equation 2.7 becomes

$$\rho = \pi an\,(n + 1)V/I \tag{2.8}$$

Usually the current input is kept constant but if b reaches such large values that the voltage between the potential probes is too small for accurate reading the current may be increased or, less satisfactorily, the value of a might be increased. It has often been argued that it is legitimate to alter a in the middle of a survey because when the Schlumberger array is used with $b \gg a$, the potential electrodes are in a region where the potential gradient is constant and therefore V/a, see Figure 2.3(B), is constant, whatever the value of a. This is true only for the case for which equation 2.7 is true - a semi infinite, isotropic, homogeneous medium.

It has also been argued that when local inhomogeneities near the potential electrodes have affected the results, this will become apparent by a displacement of the apparent resistivity curve without a change in shape. However, this implies that when a is increased, there is not a second and different inhomogeneity which affects the results in yet a different way.

Some surveys have been run with the potential electrodes initially only a few centimeters apart and buried to a depth of a few centimeters. This may violate the boundary conditions in two ways. First, the equations have been derived on the assumption that there are point sources and sinks on the surface of the earth; clearly when the potential electrode depth of burial is of the same order of magnitude as their separation, or as the thickness of the first layer, the geometry does not approximate a point source or sink. Second, in driving in the potential electrodes a local compaction inhomogeneity may be created around the electrode; if the electrodes are too close together these inhomogeneities will have a high influence on the apparent resistivity and even if the electrode separation is increased, different, but still significant, inhomogeneities may arise.

Even if there are no local inhomogeneities in the sense referred to above, there is the problem of the general inhomogeneity of a layered earth. If the first layer has a thickness which is the same order of magnitude as the potential electrode separation, then the potential field gradient in the region of the electrodes will not be uniform and increasing a in the middle of a depth sounding will give incorrect results.

This aspect of the Schlumberger array has been dealt with in some detail since many field crews have attempted to obtain detailed information on relatively thin layers (e.g. permafrost) by commencing with very small electrode separation, expanding three or four times during the survey and obtaining results which are uninterpretable.

As a general rule, if an increase of a is necessary, at least three overlapping readings should be taken, that is repeat readings for the last three values of b taken before a was increased. If the overlapped readings form a curve which is parallel to the original one it is fairly safe to attribute the difference to local inhomogeneities around the potential probes; the new portion of the curve can be displaced to fit the old one and a single interpretation applied. However, if the overlapped points do not form a parallel curve, the two sets of data should be interpreted independently.

Comparison of Wenner and Schlumberger Arrays

There have been numerous arguments as to which of the Wenner and Schlumberger arrays is better for depth sounding and profiling; much of the argument arises because of the different ways of defining depth of penetration and, in addition, confusing depth of penetration with resolving power.

If the depth of investigation (not penetration for clearly in the limit the current extends to infinity) is defined as that depth which contributes most to the total signal measured on the ground surface, then under identical experimental conditions the Wenner and Schlumberger arrays give depths of investigation of 0.11b and 0.125b respectively, where b is the current electrode separation; the system with greatest depth of investigation, 0.35b, is the two electrode array where one current electrode (S_2) and one potential electrode (P_2) are at infinity, the other current electrode (S_1) is fixed and only one electrode (P_1) is movable, b in this case being the distance between S_1 and P_1.

However, the apparently poorer depth of investigation performance of the Wenner and Schlumberger arrays is offset by their better resolving power, the Wenner array being about 10% better than the Schlumberger array and both being three to four times better than the two electrode array; here the resolving power is defined as the ability to resolve two layers vertically below one another.

That the depth of penetration of the current does not indicate the depth of investigation can easily be seen since in theory the current penetrates infinitely deep; more convincing is the demonstration, both theoretical (equations (2.4) and (2.5)) and practical, that if the current and potential electrodes of the

Wenner array are interchanged, the resistivity reading is unchanged, yet the current distributions are clearly different.

There are, however, some practical advantages and disadvantages to both methods. With the Wenner array all four electrodes have to be moved and long lengths of cables between the current electrodes and the potential electrodes may be needed; for the Schlumberger array a long length of cable is required between the current electrodes but only two electrodes have to be moved. Another advantage for the Schlumberger array is that any local inhomogeneities near the fixed voltage electrodes, including those caused by hammering in the probes, are constant during the measurements, provided the potential electrodes are not moved during the experiment. Since the spacing between the potential electrodes is usually relatively small compared with the spacing between the current electrodes higher power inputs may be needed than are needed in the Wenner array. Thus an advantage possessed by the Wenner array is that since the whole electrode array is expanded about the centre point, the fall off of potential gradient with increasing $\underline{a}$ is automatically partly compensated by the increase in potential electrode separation.

Other Array Geometries

Two other types of electrode array which are coming into favour in recent times, and used particularly with the induced polarization methods, are the dipole-dipole array, Figure 2.8(A) and the pole dipole array, Figure 2.8(B).

Strictly speaking the term dipole-dipole applies to the general case where the pair of potential electrodes may have any position and orientation with respect to the current electrodes; when all four electrodes are in line the array is usually called an axial dipole array (or sometimes polar-dipole although the term also has another meaning).

The equation for resistivity for the array shown in Figure 2.8(A) can be considerably simplified if b=a in which case the resistivity is given by

$$\rho = \frac{\pi X_1 (X_1 + a)(X_1 + 2a)}{a^2} \cdot \frac{V}{I}$$

and if $X_1 = na$

$$\rho = \pi an(n + 1)(n + 2) \qquad (2.9)$$

For the pole-dipole array the appropriate equation is

$$\rho = 2\pi an(n + 1)V/I \qquad (2.10)$$

(A)

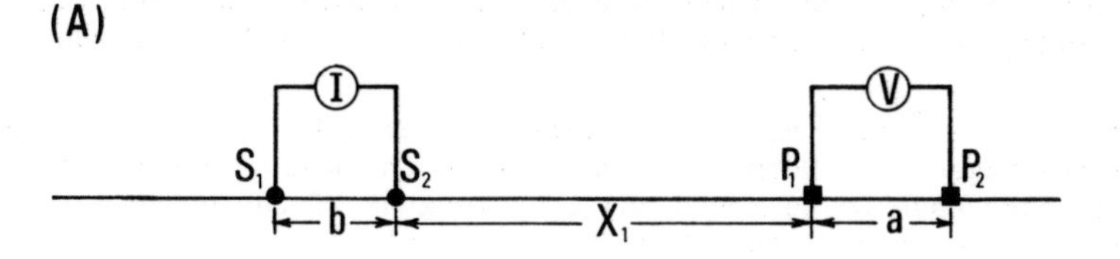

(B)

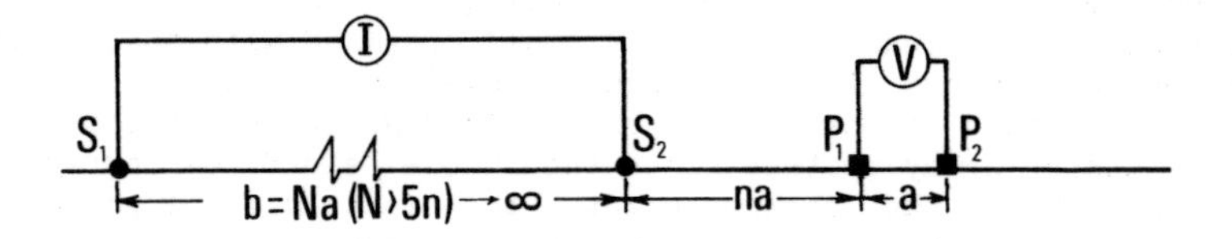

Figure 2.8. (A) Polar or axial dipole array; S_1 and S_2 are usually kept fixed, P_1 and P_2 moved in line keeping a constant and x = na. If n is large $\rho_a = \pi a n^3 V/I$. Resistivity is plotted at centre point of the array. (B) Pole-dipole array S_1 and S_2 are fixed and P_1 and P_2 moved in line keeping a constant, n is usually >5. Resistivity is plotted either at mid-point of moving current and nearest potential electrode i.e. (N+0.5n)a from S_1, or midway between moving current electrode and mid point of the two potential electrodes, i.e. (N+0.5n+0.25)a from S_1.

Using the previous definitions the depth investigation characteristics are 0.26b and 0.18b respectively, and the resolving powers are 70% and 50% respectively of the Wenner array.

The main advantage of the dipole-dipole array is the simplicity of operation since relatively short cables can be used. However, a higher power is required than in the Schlumberger or Wenner arrays.

BASICS OF CURVE FITTING

For the two layer case

$$\rho_a = \rho_1 f(k_1, \frac{a}{z_1}) \qquad (2.11)$$

where k_1 is the resistivity contrast as defined in equation

(2.19). For a given value of ρ_1, a series of curves can be computed for various values of ρ_2 and z_1. If k_1 is kept constant, that is ρ_1 and ρ_2 may vary but ρ_2/ρ_1 remains unchanged, one curve will be valid for the whole range of ρ_1 and ρ_2; since k_1 is now a constant for this curve, the apparent resistivity is no longer dependent on k_1 and equation 2.11 can be rewritten as

$$\frac{\rho_a}{\rho_1} = f(\frac{a}{z_1}) \tag{2.12}$$

Taking logarithms on both sides we have

$$\log \rho_a - \log \rho_1 = F(\log a - \log z_1) \tag{2.13}$$

which is of the form

$$y - p = F(x - q) \tag{2.14}$$

For a given depth sounding profile, values of ρ_a and a are plotted on double log paper, the curve will therefore be

$$\log \rho_a = F(\log a) \tag{2.15}$$

which is of the form

$$y = F(x) \tag{2.16}$$

It can readily be seen from equations 2.14 and 2.16 that the shapes of the two curves are identical and that only the origin is displaced. If a family of theoretical curves for various values of k_1 or, if ρ_1 has been normalised to 1, ρ_2 is plotted on double log paper, one of the curves will be identical in shape to the field curve the only effect being that the origin of the theoretical curves is displaced relative to that of the field curves; the displacement of the origin is ρ_1 in the y direction and z_1 in the x direction. Thus by curve fitting ρ_1, z_1 and ρ_2 are obtained directly, although frequently interpolation has to be made between two close fitting curves of adjacent ρ_2 values (see Figure 2.10).

For three layer (two discontinuity) cases

$$\rho_a = \rho_1 \, f(\rho_2, \rho_3, k_1, Z_1, a) \tag{2.17}$$

or

$$\frac{\rho_a}{\rho_1} = f(k_1, k_2, Z_1, \frac{a}{z_1}) \tag{2.18}$$

where

$$k_n = \frac{1 - \frac{\rho_1}{\rho_n}}{1 + \frac{\rho_1}{\rho_n}}, \qquad Z_{n-2} = \frac{z_{n-1}}{z_1} \tag{2.19}$$

and z_n and ρ_n are the thickness and resistivity of the nth layer respectively. This time a number of families of curves must be computed for various values of k_n and z_1. For each family suitable values of k_n are chosen but if ρ_1 is again normalised to 1, i.e. ρ_n is expressed in terms of ρ_1, the values of ρ_n might be listed as 1:0.2:3 this simply means ρ_2 is one fifth of ρ_1 which in turn is one third of ρ_3. Each curve in a given family will have a fixed value of Z_1; since ρ_1, k_1, k_2 and Z_1 are constant for this curve, equation 2.18 reduces to

$$\rho_a/\rho_1 = f(\frac{a}{z_1}) \tag{2.20}$$

from which we get

$$\log \rho_a - \log \rho_1 = F(\log a - \log z_1) \tag{2.21}$$

As with ρ_1, the thickness, z_1, of the first layer is often normalised to 1, i.e. thickness of the nth layer for the n layer case is expressed in terms of z_1. If any one of ρ_2, ρ_3, z_1 or z_2 are changed, the curve also changes uniquely, although it is possible that a change in, say, ρ_2 might be accompanied by a more or less compensating change in Z_1 so that two curves from different families look very similar (but are not identical).

Families of theoretical curves can be plotted on double log paper and fitted to the field curve, which has been plotted on similar double log paper, to obtain all the unknown values ρ_1, ρ_2, ρ_3, z_1 and z_2. However, the procedure takes longer then before since there are now several families of curves to sort through before a final fit is made. The four types of families for three layer cases are shown in Figure 2.9. In practice the A and Q types may be difficult to distinguish from two layer cases, depending on the actual values of k_1, k_2, and z_1.

An example of curve fitting for a three layer case is given in Figure 2.10.

For four or more layers the situation is much more difficult.

Following the style and notation of equations 2.18 and 2.19

$$\frac{\rho_a}{\rho_1} = f(k_1, k_2, k_3, Z_1, Z_2, \frac{a}{z_1}) \tag{2.22}$$

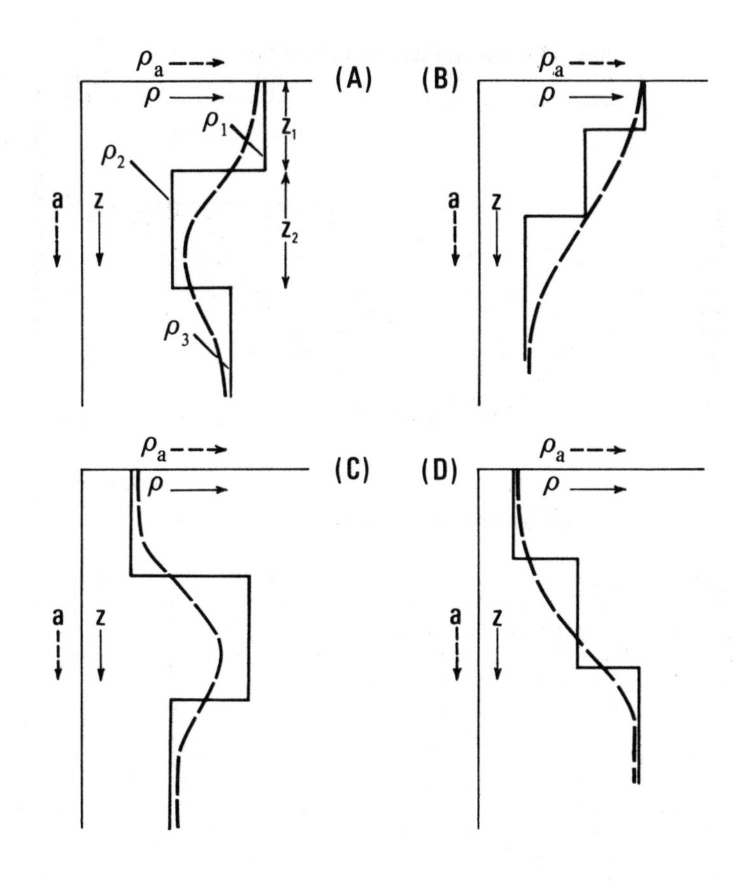

Figure 2.9. Classification of families of curves. (A) Type H. (B) Type Q. (C) Type K. (D) Type A. Solid lines represent actual resistivity structure; dashed lines represent apparent resistivity versus electrode separation for an array used in the depth sounding mode.

but there is now not only a much larger number of families of curves to consider, but also a much higher probability of obtaining very similar looking curves in more then one family by judicious simultaneous changes in k_n and Z_n. Indeed, they may closely resemble two or three layer curves as well. These comments do not alter the fact that each curve in each family is unique; the main problem is that the points on field curves are subject to scatter arising from many causes, including lateral inhomogeneities, poor and variable contact resistance at the probes, etc., and the curve is "incomplete" with respect to the theoretical curve. That is ρ_a for very large and/or very small values of a (relative to z_n) is frequently left undetermined.

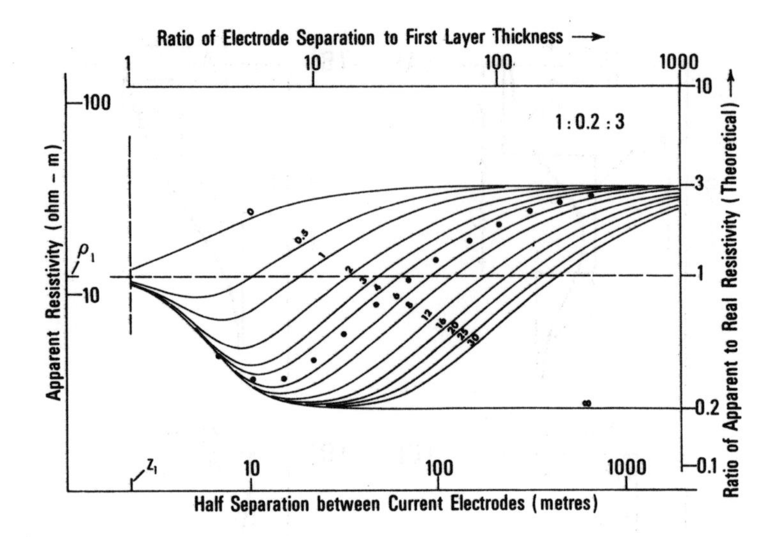

Figure 2.10. Example of curve fitting for three layers; experimental points are based upon actual measurements in the Arctic using a Schlumberger array. $\rho_2/\rho_1 = 0.2$, $\rho_3/\rho_1 = 3$. Numbers on curves are values of z_2/z_1. To obtain the field parameters the axes (dashed lines) of the theoretical curves are extended to intersect the axes (full lines) of the field curve. The points of intersection give the field values of ρ_1 and z_1. ρ_2 and ρ_3 follow from the ratios given for this family of curves. z_2 is found from the ratio number given on the best fitting curve. In this case an interpolation has been made between curves for 4 and 6. Final results therefore give $\rho_1 = 13\ \Omega$-m, $\rho_2 = 1.6\ \Omega$-m, $\rho_3 = 39\ \Omega$-m, $z_1 = 2.2$m, $z_2 = 11$m.

In general, as the number of layers involved increases it becomes more and more important to increase the range of a and number of readings taken, and to ensure parallelism of the axes when curves are being matched.

In these cases it is probably better to avoid direct curve fitting techniques and use either the auxiliary point techniques or, better still, the "compute and compare" approach.

EQUIPOTENTIAL METHOD

This is a technique similar to the point dipole technique. Two current electrodes, or a line source, are connected to the ground and are far enough apart for it to be assumed that the

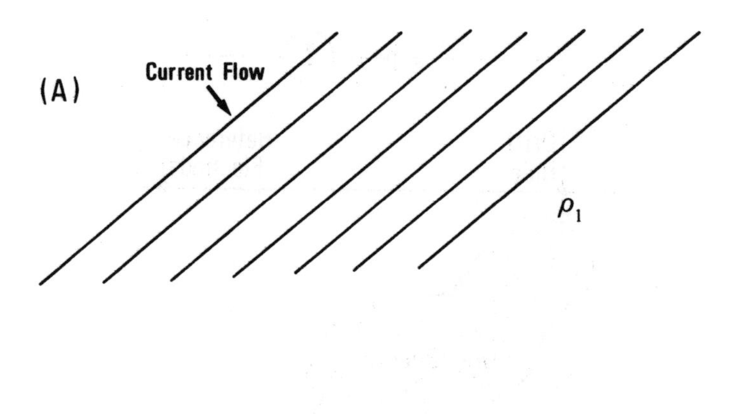

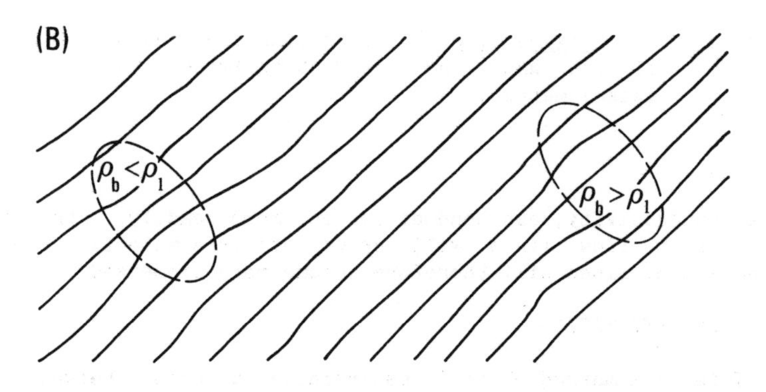

Figure 2.11. (A) Equipotential lines on surface of a semi-infinite homogeneous, isotropic medium with uniform current density source (probes at infinity). (B) Same conditions as in (A) except that bodies of different resistivities are present in the near subsurface.

current flow lines in the region being explored would be parallel to the surface of the earth, Figure 2.11(A), in the absence of an anomalous structure.

If bodies, of resistivity different from the host rock, are present in the subsurface, the equipotentials will be distorted as illustrated in Figure 2.11 (B); thus exploration of the surface using a potential electrode pair (see Figure 3.4) and contouring the results will indicate the presence of the anomaly.

Little quantitative work has been done with the method.

The simplest way to recognize which way distortion will occur is to remember that the material with the lower resistivity,

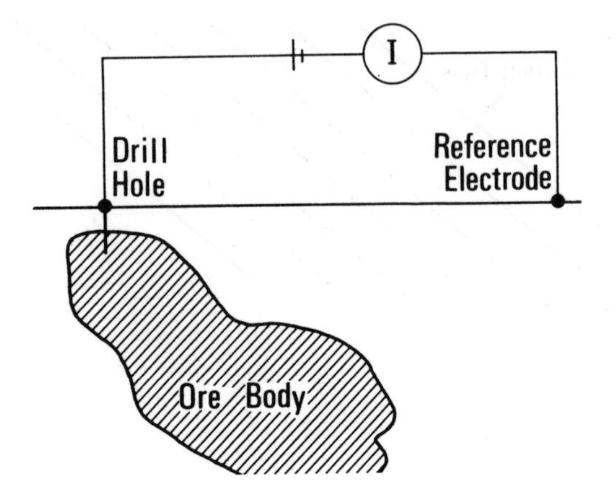

Figure 2.12. Illustrating the basis of the whole body excitation (mise a la masse) method.

higher conductivity, can conduct current more readily; therefore the current flow lines will crowd into this material, and the equipotential lines will therefore become more dispersed.

WHOLE BODY EXCITATION

This is a method which is essentially a cross between the equipotential method and the spontaneous polarization method. It is used when the orebody can itself be used as an electrode. It is most frequently employed to determine the extent of a small sulphide show in an outcrop or borehole. One of the current electrodes is plugged into the showing while the other is taken a large distance away from the showing, see Figure 2.12. Potentials are then mapped in the normal way.

Again the interpretations are largely qualitative since as can readily be seen, the similarity between this method, which uses man made potentials, and the spontaneous polarization method (Chapter 3), which uses natural potentials, means that the difficulties of idealizations are also similar.

SUGGESTIONS FOR FURTHER READING

Bhattacharya, P.K. and Patra, H.P., 1968. Direct current geoelectric sounding, Elsevier, New York.

Keller, G.V. and Frischknecht, F.C., 1966. Electrical methods in geophysical prospecting, Pergamon, New York.

Orellana, E. and Mooney, H.M., 1966. Master tables and curves for vertical electrical sounding over layered structures, Interciencia, Madrid.

Roy, A. and Apparao, A., 1971. Depth of investigation in direct current methods, Geophysics, 36, 943-959.

Saydam, A.S. and Duckworth, K., 1978. Comparison of some electrode arrays for their IP and apparent resistivity responses over a sheet like target, Geoexploration, 16, 267-289.

Tagg, G.F., 1934. Interpretations of resistivity measurements, Trans. A.I.M.M.E., 110, 135-147.

Van Nostrand, R.F. and Cook, K.L., 1966. Interpretation of resistivity data, U.S.G.S. Professional paper 499, Washington.

3

Self-potential or spontaneous polarization

INTRODUCTION

If two rods of metal are driven into the ground some meters apart and connected by a voltmeter a voltage will invariably be registered. The potentials so measured can arise from a number of causes but, ignoring man-made contributions, they may be broadly classified into two groups.

Background potentials are caused by (i) electrolytes of different concentration being in contact with one another, (ii) electrolytes flowing through a capillary or the pores of a rock (electrofiltration processes) causing what are often called "streaming potentials", (iii) contact potentials (including Zeta and Nernst potentials) at a solid-electrolyte interface, (iv) electromagnetically induced telluric currents.

These background potentials can be either positive or negative and usually range up to a magnitude of only a few tens of millivolts.

Mineralization potentials on the other hand are almost exclusively negative, range up to a magnitude of several hundred millivolts and occur principally in the vicinity of sulphide, oxide and graphite bodies.

In exploration work our interest lies mainly in the mineralization potentials with the background potentials being regarded as noise. The field measurements are easy to make both in principle and practice but quantitative interpretation is very difficult mainly because the mechanism is not fully understood.

Originally, it was believed that the observed mineralization potentials were due to oxidation of the upper part of the orebody. This has since been elegantly disproved by Sato and Mooney (1960) who have given the most useful explanation to date, their ideas being based on electrochemistry. Since their analysis of the problem is also a good example of how to assemble and interpret data their approach is closely followed here.

FIELD OBSERVATIONS

Any theory for the origin of mineralization potentials must, of course, be based on the observed field data; conversely, a successful theory must not only explain what is observed but, preferably, predict some associated phenomena which have not been observed, or which if observed had not been previously recognized as a consequence of the theory. The field data can be classified as follows.

(1) The mineral bodies which most consistently produce strong self potential anomalies are pyrite (FeS_2) and pyrrhotite (FeS). Other minerals producing less strong anomalies less consistently are chalcopyrite ($CuFeS_2$), calchocite (Cu_2S) covellite (CuS), and graphite (C).

(2) All these minerals are electronic (metallic) conductors whereas the normal country rocks in which they are imbedded are ionic (electrolytic) conductors.

(3) With the exception of graphite all the bodies are readily oxidized under easily obtained field conditions.

(4) The anomalous voltages are of the order of several hundred millivolts and are negative in the vicinity of the upper portion of the body.

(5) The observed potentials are stable with time.

(6) Only shallow ore bodies, that is those which penetrate to within 20 meters of the surface, are found by the self potential method.

(7) The region above the water table in the vicinity of bodies producing self potentials usually has a high acidity ($2<pH<5$) and an abundance of free oxygen, whereas below the water table free oxygen is almost absent and the water is slightly basic ($7<pH<9$), pH being the measure of hydrogen ion concentration in an aqueous solution.

(8) Graphite is the only non sulphide body capable of producing mineralization potentials of the same order of magnitude as those in sulphide bodies and does not readily oxidize under any normal geological field condition; whereas galena although it is a sulphide and oxidizes readily, rarely produces large self potential anomalies.

<u>Failure of Ore Body Oxidation Theory</u>

Since most self potential anomalies are found in the vicinity of bodies which are often readily oxidized, it is easy to understand why it was believed for so long that the observed potentials were a direct result of the oxidation of the ore body itself. However, by definition, oxidation is any reaction in which the valence becomes more positive; such a reaction is accompanied by the loss of electrons so we can modify the definition and say that oxidation requires the loss of electrons. A mechanism that requires the oxidation of the top portion of the ore body would imply that it was positively charged with respect to its surroundings, which is exactly the opposite of what is found in the field. Some other mechanism must therefore be formulated which, although it does not depend upon the direct oxidation of the source body, would still depend upon conditions which are favourable to some form of an oxidation - reduction system.

Basic Requirements of an Explanation

The mechanism must be capable of producing a stable electric current over large volumes for long periods of time. This implies that a great source of energy is involved since current flowing through resistive material dissipates energy.

Clearly some form of electrochemical system is required.

ANALOGIES WITH WELL KNOWN PHENOMENA

There are many occasions when chemical reactions are accompanied by the liberation of electrical energy, an electric current being produced. A practical example is the electric cell or battery, although in this case the appropriate chemical products have been formed (by electrolysis) which, under the right circumstances, have a tendency to spontaneous reaction with the liberation of electrical energy.

Since the Sato and Mooney mechanism is really nothing more than a gigantic electrochemical cell, it is worth pursuing this analogy a little further in order to understand the physical chemistry of the field situation.

Normal lead-acid batteries are called secondary cells because they are designed to be recharged - that is, by reversing the electric current passing through it we reverse the chemical reaction. A primary cell is one which is not designed and constructed for recharging after its original supply of energy has been used; no really efficient primary cell has yet been devised for high power use although the commonly used dry cell is, of course, an example of a primary cell for low power use.

One of the simplest primary cells can be formed by immersing a zinc rod and a copper or graphite rod in a dilute acid. At the negative electrode, Figure 3.1, zinc atoms give up electrons to form zinc ions which go into solution. That is

$$Z_n \rightarrow Z_n^{++} + 2\bar{e}$$

This is called a half reaction and since it involves zinc giving up electrons and going into solution it is an oxidation and anodic. The released electrons move through the external circuit to the positive electrode but they cannot move until the switch makes the circuit complete.

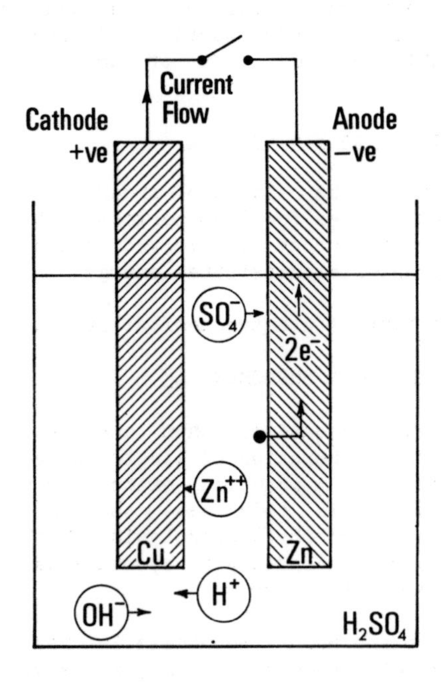

Figure 3.1. Typical electrolytic cell illustrating the principle of spontaneous polarization. Note that the terms anode and cathode are correct for a battery - i.e. a current generating device; for a current consuming device, such as electroplating cell, electron gun, etc., the terms are frequently reversed.

At the positive electrode (cathode) the electrons are picked up by the hydrogen ions in the acid to form hydrogen atoms and, in turn, hydrogen molecules; in this particular cell the cathode eventually becomes coated with hydrogen gas and the reaction stops. In other words

$$2H^+ + 2\bar{e} \rightarrow 2H \rightarrow H_2(\text{gas})$$

and this is also called a half reaction, is a reduction and cathodic.

In the electrolyte the current is carried between the electrodes by the ions in solution, remembering that by convention the current flow is in the same direction as the positive ion flow but in the opposite direction to electron flow or negative ion flow.

In this cell the two chemicals zinc and hydrogen constitute what is called an electrochemical couple (i.e. a zinc-hydrogen ion couple). The reaction occurring between the electrochemical couple in the cell is called an oxidation-reduction reaction and can be written simply

$$Zn + 2H^+ = Zn^{++} + H_2$$

The electron releasing or attracting power of ions (or atoms) in solution can be listed in a table called the electromotive force series or standard electrode potentials. The term oxidation potential (Eh) is essentially the same thing but the potential is of the opposite sign; sometimes the term oxidation potential is used interchangeably with the term redox potential.

It is not possible to measure the voltage of a half reaction alone - there must be two half reactions. For the purposes of the electromotive force series the voltage of the hydrogen half reaction is arbitrarily set to zero and the voltages of other elements in ionic solutions are referred to this.

In general, the electrochemical reaction in aqueous media can be given in the form, following Sato and Mooney, of

$$aA + wH_2O = bB + m\,H^+ + ne^{-1}$$

where _a_, _w_, and _b_ are the mole numbers (gram molecular weights) of the participating substances, _m_ is the hydrogen ion concentration (pH) which is a measure of the influence of the hydrogen ions, and _n_ is the oxidation potential (Eh) which is a measure of the influence of the electrons. Conditions for thermodynamic equilibrium can be expressed in terms of the two independent variables _m_ and _n_; any specific reaction could be independent of pH and Eh in which case _m_ and _n_ are zero and we have a chemical reaction without the release of electrical energy.

If the second law of thermodynamics is now applied to any system it is found that the potential generated depends upon the activities (approximately the concentrations) of the reactant and product as well as on Eh and pH.

GENERAL FIELD MECHANISM

It is found that the potential difference which produces the currents is due to the difference in the oxidation potentials of the solutions immediately in contact with the orebody at its upper and lower end. Near the upper end the substances and solution are in relatively oxidized states while near the lower end they are in relatively reduced states. The half cell reaction at the upper end is therefore one where the oxidation-reduction reactions take place between the relatively oxidized state of the country rock reducing material at the surface of the orebody; at the lower end the relatively reduced states of the country rock react to oxidize material at the surface of the body at the lower end. The oxidation process at the surface of the lower end of the orebody releases electrons which then travel through the orebody to the upper surface, Figure 3.2, the electrons then being donated to the material at the upper surface of the orebody as it is being reduced by the oxidized states of the country rock. The current flow through the orebody is therefore from top to bottom (that is in the reverse direction to electron flow) the return current in the country rock being by motion of the ions through the pore

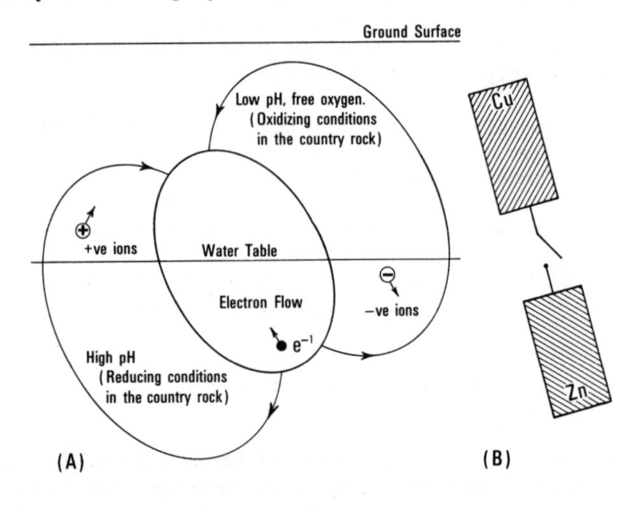

Figure 3.2. (A) Illustrating a typical geological situation and current flow system for spontaneous polarization. (B) By rotating Figure 3.1 clockwise and unwinding the metal rods, the analogy with the geological situation is obvious.

fluids; the positive ions travelling from the lower end to the upper end and the negative ions from the upper end to the lower end.

The orebody itself does not participate directly in the electrochemical reactions producing the self potentials; in this sense the orebody acts as a chemically inert medium, its only function being to transport electrons.

Thus we have essentially two separate half cell reactions, one anodic and one cathodic which could be chemically unrelated to one another and may go on regardless of whether or not an orebody is present. However, once they are coupled through a medium of electronic conduction, such as an orebody, spontaneous polarization may take place giving the observed self potential anomalies. Independently the two half cell reactions may produce electrons at different rates but once connected they become self balancing since the reaction with the lowest rate limits the other one. In principle, the orebody need not be partially immersed in the water table although in practice such a condition increases the probability of the appropriate electrochemical couple being produced.

The specific electrochemical reactions concerned in any given field condition are still not well understood. However, the source of energy, which is required to maintain reactions over long periods of time is essentially oxygen in the air. This is always in contact with the ground so that free oxygen is readily dissolved in the solution surrounding the orebody and the oxygen can then combine with hydrogen or hydroxyl ions to form hydrogen peroxide, which in turn can decompose to form hydroxyl ions and free oxygen. The only material consumed is oxygen which can be replaced from the air.

The work of Sato and Mooney considers a number of possible specific reactions of which the most probable require the presence of iron, both Ferrous and Ferric (Fe^{++} and Fe^{+++}), and manganese, both of which are quite ubiquitous in nature, (Figure 3.3).

<u>Some Geologic Implications</u>

1. In order to provide the appropriate environment to transfer electrons from the region of lower oxidation potential at its lower end, the orebody, must be a good electronic conductor. In other words it must provide a continuous connection between groundwaters of different oxidation potentials.

2. The ore must be chemically stable or inactive with respect to the oxidation potentials of the solutions with which it is in contact otherwise it will gradually disappear with time and such bodies would not be observed; many of the orebodies are known to be of Precambrian age implying highly stable and inert material.

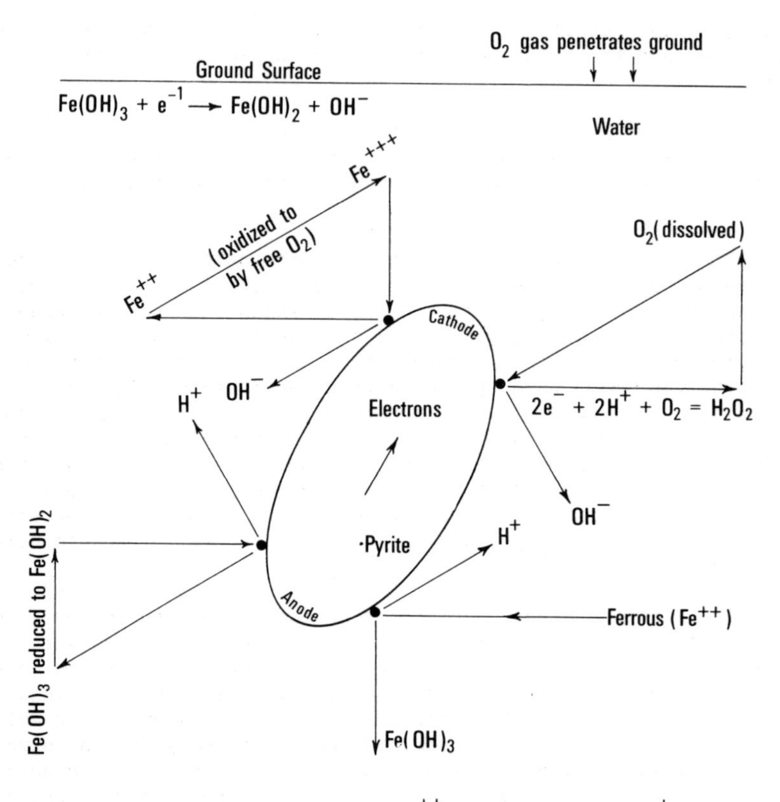

Figure 3.3. One suggested detailed mechanism which gives spontaneous polarization potentials of the right order of magnitude under typical field conditions (after Sato and Mooney 1960).

3. Insulating films of a permanent nature cannot form on the surface of the ore, or if they do the orebody is removed from the electrochemical system since a transfer of electrons cannot take place at the surface, and spontaneous polarization potentials disappear.

4. The magnitude of the spontaneous polarization potential will rapidly diminish with ore grade since the electronic conductivity decreases with decreasing ore grade.

5. In arid country, or in areas of permafrost, the conditions are not favourable for spontaneous polarization potentials because of lack of oxygen in solution to support the half cell reactions, and lack of solutions to carry charged ions.

6. The apparently anomalous position of galena, which does not often produce significant spontaneous polarization potential anomalies, is readily explained since it frequently occurs as separate lenses with no electrical connection between them; in addition, an insulating coating of lead sulphate is often formed on its surface.

Self Potentials in Geothermal Areas

There appears to be only one confirmed source of high self potentials which do not originate from spontaneous polarization. These occur in geothermal areas where high conductivity fluids undergo strong convection. The mechanism of the origin of the potentials (as high as 1 volt) is not completely understood but they may be a special case of streaming potentials; alternatively they may be simply an example of a current system being augmented due to the mass transport of electrons with the highly mobile conducting fluid.

FIELD MAPPING OF POTENTIALS

The field procedure for the mapping of spontaneous polarization potentials is relatively simple, Figure 3.4. Two electrodes are required between which is connected a sensitive high impedence d.c. voltmeter. Special care must be taken with the form of the electrodes since without this care strong

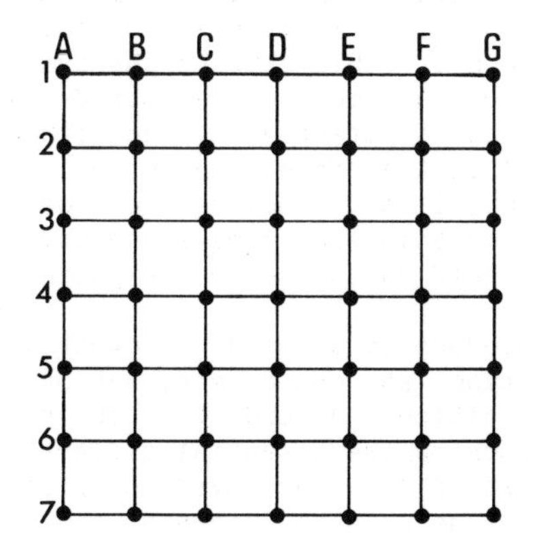

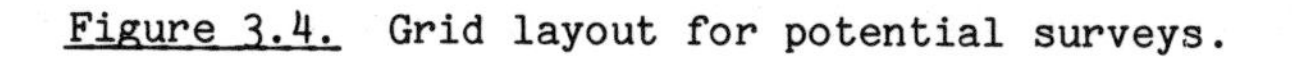
Figure 3.4. Grid layout for potential surveys.

electrode potentials may be set up which interfere with the natural potentials. A number of types of electrodes have been designed, all requiring a saturated solution of a metal salt. One of the simplest, convenient to make and in wide use, consists of inserting a coppper rod in a saturated solution of copper sulphate, the copper sulphate solution being contained in a porous pot. It is important that the solution be saturated since in a dilute electrolyte dissociation occurs so that ions have high mobility and readily move to the electrodes when an electric current is passed through the solution. As the electrolyte is made stronger the "activity" and mobility decrease until with a saturated solution the ion pairs cannot be readily polarized.

Contact is made with the ground which may need to be wetted with a bucket of water if it is too dry. This means that there may still be a spontaneous polarization contribution due to the contact between two electrolytes of different concentrations - namely, that of the copper sulphate solution in the pores of the pot and the electrolyte in the ground on which the porous pot is embedded; however, this contact potential is not likely to exceed a few millivolts, compared with the few hundred to several hundred millivolts of the anomaly being sought and the contact potential will simply constitute noise.

The area of interest is divided into a square grid with, say, 10 to 50 meter square sides, and measurements are made using one of two procedures.

The first procedure consists of placing one of the electrodes at a large distance (infinity) away from the anomalous area. This electrode gives a reference potential against which the potential at each of the corners of the square grid is measured. The results are then contoured or a series of profiles produced in the normal manner. The chief disadvantage of this method is that it requires a long length of cable between the two electrodes; the chief advantages are that only one electrode is physically moved in the field, any contact potential at the other electrode (the reference electrode) is therefore constant so that only variations in the contact potential at the moving electrode are a source of noise. A variant in this procedure is to use the potential at a point on the grid itself as the reference potential.

The second procedure is to measure potential differences on the grid by starting at one corner, say by placing one of the electrodes, P_1 at position A1 and the other electrode P_2 at position B1 and measuring the potential difference between the two as ΔV_B; the electrode P_1 is then moved to position C3 and the potential difference between positions B1 and C1 measured as ΔV_C, care being taken to note the sign change in the voltage because of the essentially reversed electrodes compared with the first measurement. The whole grid is covered in this way. The results may be contoured or profiled as field gradients but since the results are difficult to interpret even in a qualitative manner, it is more usual to take one of the points as a reference and plot all other potentials relative to it. The chief disadvantage of this method is the extra work involved in summing the potential differences before contouring or profiling; the principal advantage is the short length of cable required in the field. An examination of the method of adding the potential differences between pairs of points will show that the process of "leapfrogging" the electrodes also minimizes the contact potential noise problem which might be suspected because of the movement of both electrodes.

Under highly idealized field conditions it is possible to obtain some information about the size of the body; there is little point in obtaining information about depth since, as indicated in the discussion about the mechanism, only very shallow bodies are likely to give rise to a spontaneous polarization potential. However, so ideal must the subsurface conditions be to make the theoretical approach tractable, nearly all interpretations are qualitative. They consist essentially of noting where the minimum in the anomaly lies, determining if there is any significant asymmetry in the contours and drilling a shallow hole (not greater than 50 to 100 meters deep), collared at or near the anomaly minimum and angled if necessary, to take account of the asymmetry in the anomaly contour. It is more usual to use further simple geophysical techniques, e.g. a magnetic survey, before making a decision to drill.

SUGGESTIONS FOR FURTHER READING

Anderson, L.A. and Johnson, G.R., 1976. Application of the self potential method to geothermal exploration in Long Valley, California, J. Geophys. Res. 81, 1527-1552.

Corwin, R.F. and Hoover, D.B., 1979. The self-potential method in geothermal exploration, Geophysics, 44, 226-245.

Petrovski, A., 1928. The problem of the hidden polarized sphere, Phil. Mag. 5, pp. 334-353 and 914-933.

Sato, M. and Mooney, H.M., 1960. The electrochemical mechanism of sulphide self potentials, Geophysics, 25, 226-249.

Yungal, S., 1950. Interpretation of spontaneous polarization anomalies caused by spheroidal ore bodies, Geophysics, 15, 237-246.

4

Induced polarization

In the chapter on resistivity methods one of the implicit assumptions made was that upon switching on the current between two current electrodes the voltage measured between the two potential electrodes was produced instantaneously. In general, this is not correct since the rise time of the voltage is finite. The delay in the voltage reaching its maximum value arises from a number of causes which may be broadly classified as instrumental effects and geological effects. The instrumental contribution to the delay time can be quite small and in those areas where the simple resistivity method is applicable the contribution to the delay time from subsurface geological causes is also too small to be of much use. However, there are some types of geological situations where the rise time is significant; this effect varies from place to place so the length of the rise time and the shape of the curve may therefore constitute a useful diagnostic method for exploration.

The areas where the induced polarization method has been most useful are those where disseminated sulphide deposits occur, such as the porphyry type, bedded copper deposits, bedded lead-zinc deposits in carbonate rocks and, less commonly, in some types of clay deposits. There are two principal causes of induced polarization effects - electrode polarization and membrane polarization.

ORIGIN

Electrode polarization

To understand the processes it is useful to discuss the analogy of the electrochemical cell introduced in the section on spontaneous polarization.

If a metal is immersed in a solution of ions of a certain concentration and valence, a separation of charge is found to occur and a potential difference is established between the metal and the solution, Figure 4.1.

When an external voltage is applied across the interface the ionic equilibrium is disturbed causing a current to flow and the potential difference across the interface changes from its initial

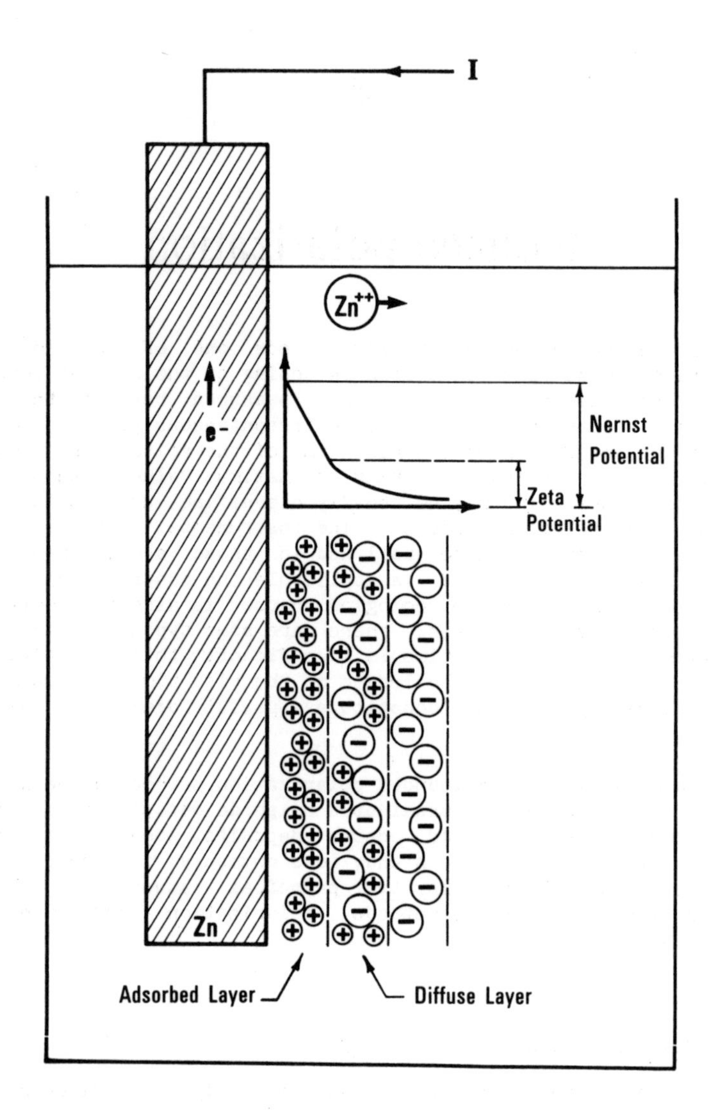

Figure 4.1. Illustration of the elements of electrode polarization.

value; when the external voltage is removed the initial voltage is gradually re-established.

In the geological environment it is useful to consider the problem using as an example the structure of a medium of disseminated sulphides, the most fruitful use for the induced polarization method. Such a structure may be regarded as metallic

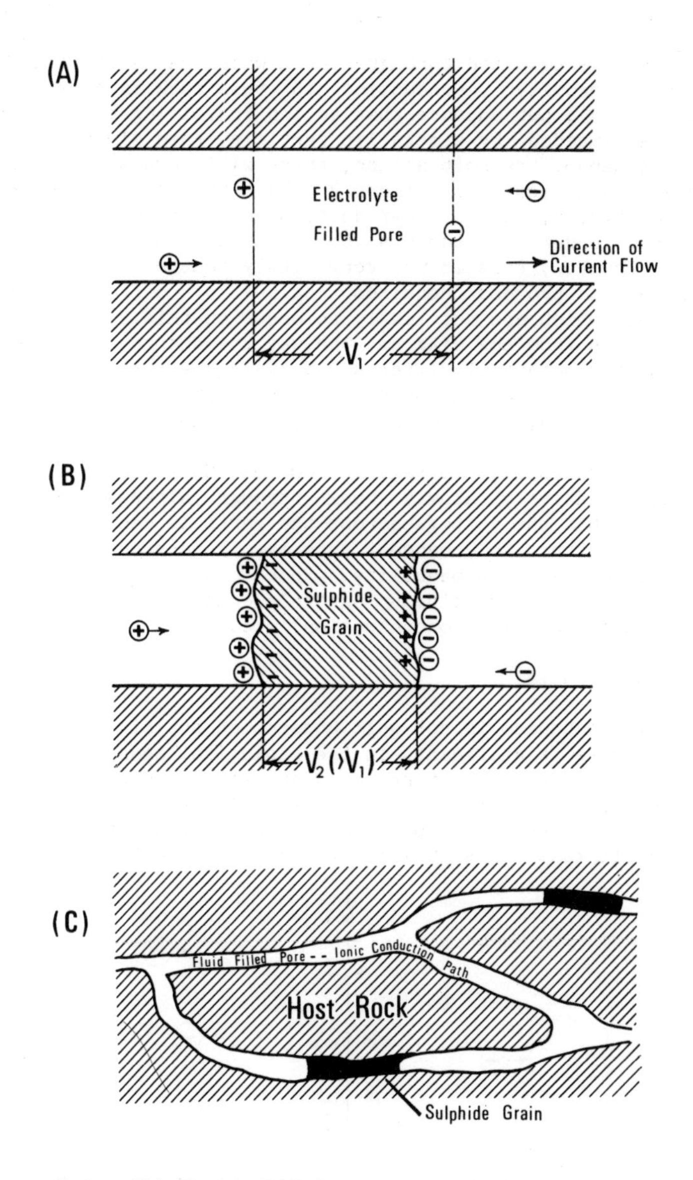

Figure 4.2. (A) Fluid filled pore in a rock indicating potential, V_1, which would be measured between two planes in conditions of free ionic motion. (B) Same pore as in (A) but blocked by a sulphide particle between the planes in (A). Ionic charges build up, a new potential, $V_2(>V_1)$, would be measured, and "overvoltage" developes. (C) General scheme for alternate ionic paths when some pores are blocked by sulphide particles.

particles disseminated through a rock which has a certain fracture or void porosity. Conduction in this rock is normally ionic, Figure 4.2(A), but suppose that an ion travelling through the rock pore is impeded by a metallic particle, then since the metallic particle cannot transmit an ion, there will be a pile-up of ionic charges as indicated in Figure 4.2(B). Some of these charges may be discharged by oxidation-reduction couples similar to those discussed under spontaneous polarization but on a much smaller scale. However, this is not very likely because the spontaneous polarization process requires regions of different oxidation potentials which are not likely to occur on this fine a scale. As can be seen from the diagram, the way the charges pile up tend to increase the potential difference between the exterior surfaces of the particle - hence the phenomenon is often known as overvoltage; the particles may also be said to be polarized.

When the applied voltage is removed the piled up ions tend to diffuse relatively slowly back through alternate paths in the medium so that the voltage across the particle decays to zero in a finite time that is dependent upon may factors such as the actual rock structure, permeability, pore diameter, electrolytic conductivity of pore fluids, particle conductivity (and therefore by inference metallic concentration) etc.

Membrane Potential

This mechanism is not well understood but since the potential only becomes significant in the geological sense when there are clay minerals present, it probably arises through a process of ion sorting such as electro dialysis. In this process ions of different diameter are separated out due to fineness of the pores and capillaries; this is illustrated in Figure 4.3.

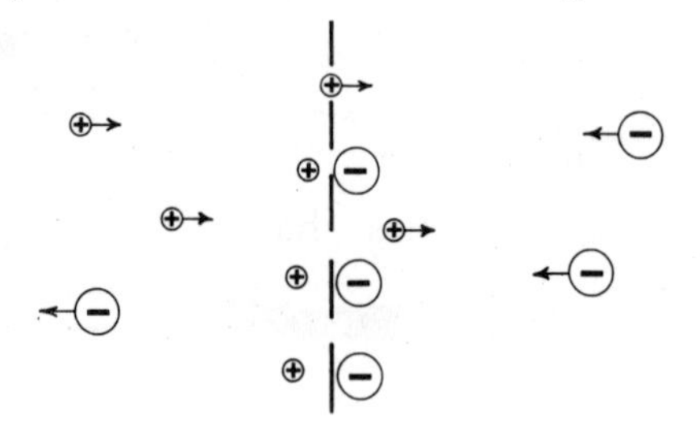

Figure 4.3. Illustration of how electrical double layers arise, causing "membrane potentials".

The clays act as a very fine filter, the pore diameters being of the same order as the ion diameters so that as the ions move the negative ions, which are generally larger than the positive ions, get held up at the filter while the positive ions can pass through, thus leading to a negative surface charge. These negative charges attract a certain number of positive charges to them in order to maintain electrical neutrality and an electrical double layer results.

When an external potential difference is applied in the region, that is when an electric current is forced through the clay, the electrical double layer equilibrium is disturbed and a new form of equilibrium is established over a certain period of time; when the external current is removed the charges tend to re-distribute themselves to their former equilibrium pattern but since this takes a finite period of time an induced polarization effect is observed.

Common Cause of Electrode and Membrane Potentials

The most important point to be noticed is that both the electrode and membrane polarization effects seem to depend on blocking the natural passage of ions in an ionic conductor (electrolyte).

In the geological environment, membrane polarization is regarded as a nuisance effect because it complicates the measurement and interpretation of IP observations when prospecting for minerals; there is very little use for the IP method in detecting clays. We will therefore consider only electrode polarization effects in more detail.

CONCEPT OF CHARGEABILITY

If in a homogeneous medium we introduce current at two or more points (sources and sinks) then the primary potential drop between two other electrodes serving as potential electrodes will, in the absence of volume polarization effects, be given by

$$V = \frac{I}{\sigma} \cdot F(s,g) \qquad (4.1)$$

where F(s,g) is a function of the <u>s</u>ize and <u>s</u>hape of the body, and of the <u>g</u>eometry of the various electrodes, and σ is the electrical conductivity of the medium. It may be recalled from Chapter 2 that for the Wenner array F(s,g) is equal to $(2\pi a)^{-1}$ where <u>a</u> is the electrode separation.

Whatever the actual physical mechanism, when polarization occurs we can represent it by a volume distribution of electric dipoles, rather similar to miniature battery cells, whose strength is proportional to the primary current density, say J, and which opposes the primary current, see Figure 4.4. If the constant of proportionality is m then when we have polarization effects the current density will be $J - m J = J (1-m)$.

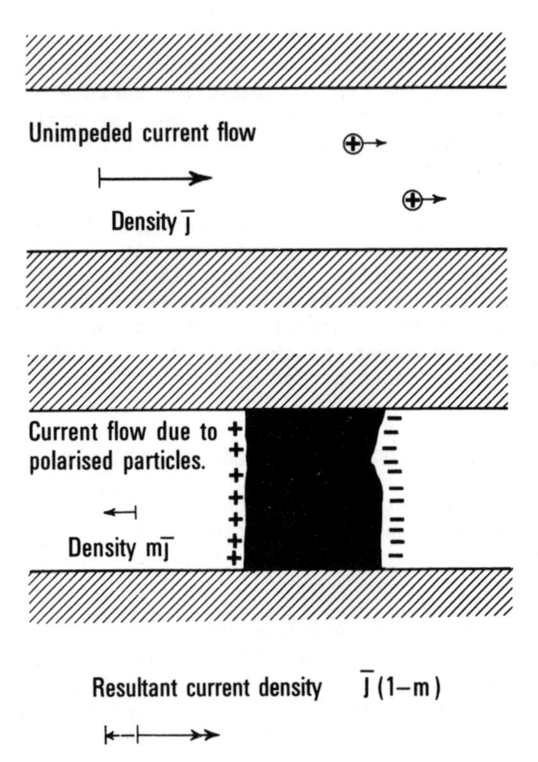

Figure 4.4. Showing how, given a fixed power source at the surface, electrode polarization effects cause a reduced current density underground compared with the current density which would arise in the absence of polarization.

In relating this to the determination of the underground physical property from surface measurements, the reduction in current density will appear as a reduction of m in the conductivity of the medium. That is, in the presence of polarization the potential drop between the voltage electrodes will be increased to

$$V_o = \frac{I}{\sigma(1-m)} \cdot F(s,g) \qquad (4.2)$$

The increase in voltage, i.e. the secondary voltage V_s, is given by

$$V_s = V_o - V = I \, . \, \frac{F(s,g)}{\sigma} \frac{m}{1-m} \qquad (4.3)$$

and the ratio of peak secondary voltage to the observed steady state voltage is

$$V_s/V_o = (V_o - V)/V_o = m \quad (4.4)$$

We now call m the chargeability of the medium. The expression for m contains no geometrical factor so that ideally, true chargeability is a volume effect. Thus the ratio V_s/V_o should be independent of topography and, for homogeneous isotropic samples (as used in the laboratory, for instance) independent of the electrode geometry, specimen size and shape. There are also some subsidiary effects due to temperature variations and electrolyte concentrations.

The question now is, what does m mean in terms of what is measured in the field. This is perhaps best illustrated by considering what is known as the Pulse-Transient (or Time Domain) method.

PULSE-TRANSIENT METHOD

Application of a direct current at some instant in time, with its subsequent removal, will result in a voltage-time curve, similar to that shown in Figure 4.5. The observed voltage, V_o, is that which we actually observe due to the applied current plus polarization effects. The peak secondary voltage, the V_s, is that which we would observe immediately we remove the applied direct current after applying it long enough to reach a steady voltage. From Figure 4.5 it can easily be seen $V_s = V_o - V$.

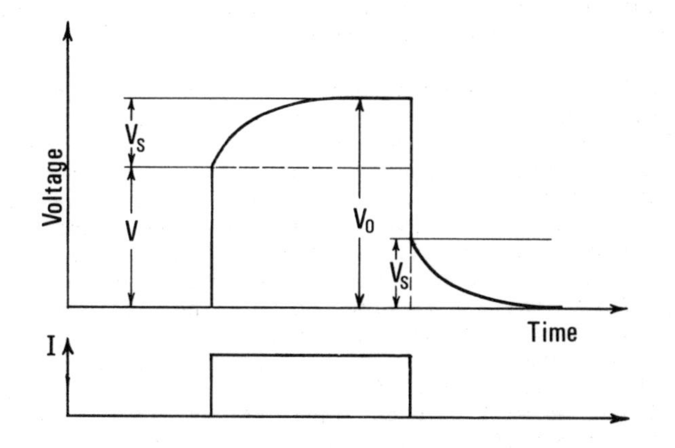

Figure 4.5. Comparison of current-time and voltage-time curves in the presence of IP effects. Dashed lines represent voltage-time curve in absence of IP and induction.

In actual field measurements the true chargeability of the medium cannot be measured - or if it can it is of no interest since it would imply that the subsurface is homogeneous. The search is for regions of anomalous chargeability. Hence what we really measure is the apparent chargeability m_a which is a complicated function of the true chargeabilities and resistivities of the various media within the range of the equipment; the actual function does not concern us at this level.

It is vitally important to note that when we talk about m_a it is apparent chargeability on three counts.

(a) It is a function of all m_i for i layers
(b) V and other voltages depend on the current charging time, unless t goes to infinity
(c) Because of practical problems V_s cannot be measured, so that some other voltage $(<V_s)$ is used.

The practical problems in trying to observe V_s directly arise principally because of its transient nature. The actual process of switching the current on and off instantaneously results in secondary transient voltages, both in the geological medium and in equipment. This occurs in any system that has inductance. We therefore have to wait for a little while for the transient (which must be of much higher frequency than the IP diffusion effects) to die away and/or arrange for the coupling to produce insignificant magnitude in the secondary transients. These points are of more importance in another method known as the "frequency effect method" and will be discussed later. Even for a perfectly homogeneous medium we can only measure the voltage a short time after the applied current is removed.

In principle, we could obtain a complete record of the voltage-time curve, after allowing transients to die down, and extrapolate (or curve fit) back to the instant of cut off to obtain V_s directly. However, until recently there were many practical difficulties against this, the principal ones being the considerable amount of extra field equipment and field time required as well as computational time in the laboratory. Thus we do not measure V_s but some other voltage which may be expressed as $V_s .F(m_a, V_o, t)$ where t is time and the other letters have their previous meanings. Since V_o can be measured, and the function $F(m_a, V_o, t)$ depends mainly on m_a, if we always measure the polarization voltage at a given time after the impressed current is turned off, we will have an idea of at least the relative values of V_s for different apparent chargeabilities m .

If V can be adjusted so that V_o is always the same for every experiment, then the measure of variations in polarization voltage V_s, after allowing sufficient time for transients to die away, is also a measure of variations in m_a.

To simplify matters in the field the area under the voltage decay curve is integrated over small intervals of time (usually of the order of a half second) as soon after cut-off as possible (again after about a half second), see Figure 4.6; this area has units of volts-seconds. However, as the input voltage increases so does the secondary voltage, so the integral is normalised by dividing it by V_o; the units in this case are seconds, although milliseconds are often used so as to give reasonable numbers. The values obtained are plotted against field position. If V_o is kept constant, as can readily be done by black box gadgetry, then this voltage need not be measured separately.

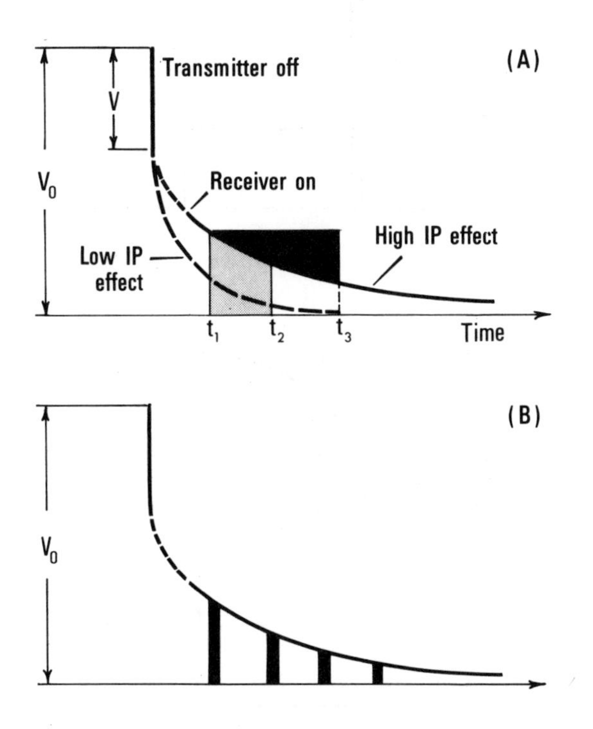

Figure 4.6. (A) Typical procedure for pulse transient method. Off period allows high frequency transients to die down before switching on receiver; t_1, (t_2-t_1) and (t_3-t_2) are usually of the order 0.5 secs. Area under curve between t_1 and t_2 is integrated and often is normalised to V_o. Second area above curve is sometimes integrated to give more information on shape of decay curve. (B) Alternative approach to obtaining more information on shape of decay curve; time interval for integration, $\Delta t \simeq 0.1$ sec.

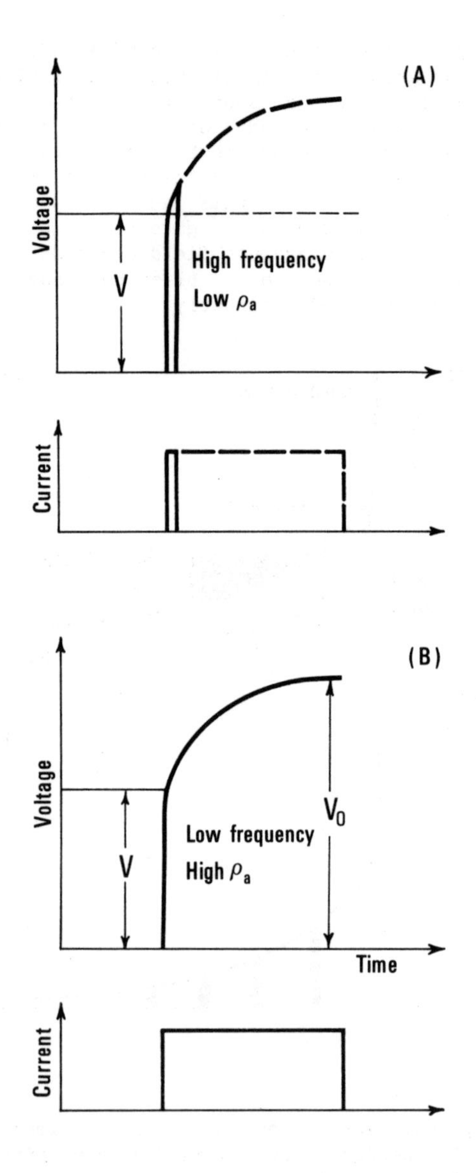

Figure 4.7. Illustrating why different apparent resistivities are obtained at different frequencies when IP effects are present. In (A) the value of ρ_a obtained is the same as that which would be measured in the absence of IP effects.

The principal advantage of integrating and averaging is that the process suppresses earth noise and cross coupling effects between cables. Obviously, care has to be exercised in the choice of a suitable time interval since it should be long enough to maximise the noise reduction but not so long that the method becomes diagnostically insensitive.

In this context, one of the most useful diagnostic properties is the complete curve shape. With recent improvements in technology, it has become possible to integrate over relatively small time intervals without significant error from noise. Then if the time interval of integration is small enough, for example Δt less than 100 msec, and the integral is divided by the time interval Δt, a good approximation of the average voltage over that interval is obtained. If this is done a number of times along the decay curve see Figure 4.6(B), a reasonable approximation to the complete curve can be reconstructed.

VARIABLE FREQUENCY METHOD

In this method, also known as the Frequency Domain or Frequency Effect method, the principal interest lies in the change of apparent resistivity with change in the frequency of the applied current. That the apparent resistivity changes as the frequency changes can be seen from equations 4.1 and 4.2 where

$$V = \frac{I}{\sigma} . F(s,g) = I . \rho . F(s,g)$$

and

$$V_o = \frac{I}{\sigma(1-m)} . F(s,g) = I . \rho_a . F(s,g)$$

If a measurement of apparent resistivity is made using a high frequency in, for instance, a Wenner array, then, referring to Figure 4.7(A) it can be seen that there is only time to measure V as a voltage, since the V_s has not had time to build to a maximum; therefore, at high frequency we have low voltage, high apparent conductivity and low apparent resistivity. The resistivity measured at high frequency is clearly that which would be measured if there were no IP effects present.

On the other hand, referring to 4.7(B), at low frequency the voltage measured has had time to build to its maximum, V_o, so that at low frequency we have high voltage, low apparent conductivity, and high apparent resistivity. That is as frequency, f, goes up the apparent resistivity goes down.

<u>Definition of Frequency Effect</u>

To put resistivity information to good use a parameter called the "frequency effect", frequently shortened to FE, is defined as

$$FE = \frac{\rho_0 - \rho_\infty}{\rho_\infty} \qquad (4.5)$$

where ρ_0 is the apparent resistivity at zero frequency, i.e. using a direct current, and ρ_∞ is the apparent resistivity at infinite or very high frequency. In practice it is found that f = 0.05 Hz is close enough to DC, and f = 10 Hz is close enough to ∞ to give a reliable FE value. It is clear then that the method is based on the determination of the apparent resistivities of the subsurface at high and low frequencies.

In practice, it might be found that f = 0.05 Hz and f = 10 Hz is too wide a variation, principally because use of such a low frequency means that much more field time is required for a single measurement than would be the case if the frequency was, say, 0.1 - 0.2 Hz. The proper choice of frequencies depends on many things, not the least of which is the time allowed and the cost of the field work. Thus, in some surveys use of f = 0.05 Hz and 10 Hz might take three or four times as long as a survey using f = 0.3 Hz and 5 Hz with little or no gain in additional information. Some preliminary experimentation and experience soon leads to a reasonable compromise.

RELATION BETWEEN CHARGEABILITY AND FREQUENCY EFFECT

In the pulse transient method reference was made to the conductivity of a subsurface medium whereas in the variable frequency method the inverse, the resistivity, is used; this change in terminology is unfortunate but is retained here since both are widely used. However, it is obvious that the terms refer to the same physical property and that there should therefore be a simple connection between "chargeability" and "frequency effect".

Equation 4.1 gives the voltage-current relation observed in the absence of IP effects; clearly the σ is equivalent to the conductivity found at high frequency or, more specifically,

$$V = \frac{I}{\sigma_\infty} F(s,g) \qquad (4.6)$$

In the presence of IP effects it was argued that on the surface it would appear as if the conductivity had decreased and a reduction factor of (1-m) was used where m was defined as the chargeability. That is

$$V_o = \frac{I}{\sigma_\infty(1-m)} F(s,g) \qquad (4.7)$$

But in fact, the conductivity so measured would be the same as that measured using DC in the variable frequency method. That is

$$V_o = \frac{I}{\sigma_o} F(s,g) \qquad (4.8)$$

From 4.7 and 4.8 $\sigma_o = \sigma_\infty (1-m)$ or

$$\rho_\infty = \rho_o (1-m) \qquad (4.9)$$

From the defining equation 4.5 the frequency effect can be rewritten as

$$FE = \frac{\rho_o - \rho_o(1-m)}{\rho_o(1-m)} = \frac{m}{1-m} \qquad (4.10)$$

or

$$m = \frac{FE}{1 + FE} \qquad (4.11)$$

If FE<<1, that is, if ρ_o is approximately equal to ρ_∞ implying that the medium is weakly polarizable, then m = FE.

Equations 4.10 and 4.11 show that the two methods are essentially measuring the same thing; either of them can yield the Frequency Effect or the chargeability, and it is simply a matter of convenience which method is used.

METAL FACTOR

A further development often used with the variable frequency method but which, because of the previously discussed relation can also be used with the pulse transient method, is the introduction of a quantity called the metal factor, MF, or metallic conduction factor (MCF).

This is obtained by dividing the frequency effect by the DC resistivity and multiplying by $2\pi x10^5$, the constant multiplying factor simplifies calculation and gives convenient numbers; the 2π arises because of the assumed semi-infinite geometry (see equation 2.4) and the 10^5 gives numbers which avoid the need of continually writing down a very small decimal fraction e.g. 351 instead of 0.00351. Therefore

$$MF = \frac{FE}{\rho_o} \times 2\pi \times 10^5 = \sigma_o \times FE \times 2\pi \times 10^5 \qquad (4.12)$$

The principal advantage in using the metal factor is that it tends to compensate for the higher conductivity of obviously conductive, but non IP, environment such as concentrated sulphide deposits or disseminated sulphides in shear zones; the sulphides would usually have better than 1 % concentration (but less than 30% concentration since these ore bodies are usually more quickly and cheaply discovered by electromagnetic methods).

The reasons for this advantage can best be seen by recalling the definition of frequency effect and recasting equation 4.12 into the form

$$MF = (\sigma_{\infty} - \sigma_{o})\ 2\pi \times 10^{5} \tag{4.13}$$

From this it can be seen that the metal factor is essentially a measure of the change of apparent conductivity with change in frequency. If the conductivity doesn`t change much with frequency, i.e. if there are no significant IP effects, then the metal factor will tend to zero, whereas it does not when IP effects are present. Therefore, even in conducting environments the metal factor will be zero if there are no IP effects.

Complex Resistivity Measurements (CR)

One of the most recent areas of interest in the frequency domain, but not yet a thoroughly field tested and generally accepted technique, is the complex resistivity method. Because there is a finite rise time to the voltage being measured after the current is switched on or off, it is obvious that the voltage in the ground is not in phase with the primary current. The phase difference, or phase angle, may be just as diagnostic of the subsurface origin as the voltage decay curves especially as the frequency dependence of the phase angle and voltage may be different from each other as well as different for each class of mineral deposit. For example, a recent experiment differentiated between graphite, where the phase angle increases with frequency, and massive sulphides where the phase angle decreases with frequency.

PLOTTING OF RESULTS

Any of the following parameters can be used when the results of a field survey are to be presented

(a) apparent chargeability,
(b) apparent ρ_o
(c) FE
(d) MF

All can be plotted against field positions at which the results were obtained, but traditionally there are differences of approach between the Time Domain and Frequency Domain methods. In the Time Domain method the chargeability is plotted against field position to produce a profile.

For the Frequency Domain method, it is more usual to plot pseudo sections for presentations of ρ_o and either FE or MF for each survey rather than rely on only one parameter. However, because of the relationship between apparent chargeability and the other parameters, there is no reason why pseudo sections cannot be plotted for m_a also when the Time Domain method is used.

The method of plotting pseudo sections is shown in Figure 4.8.

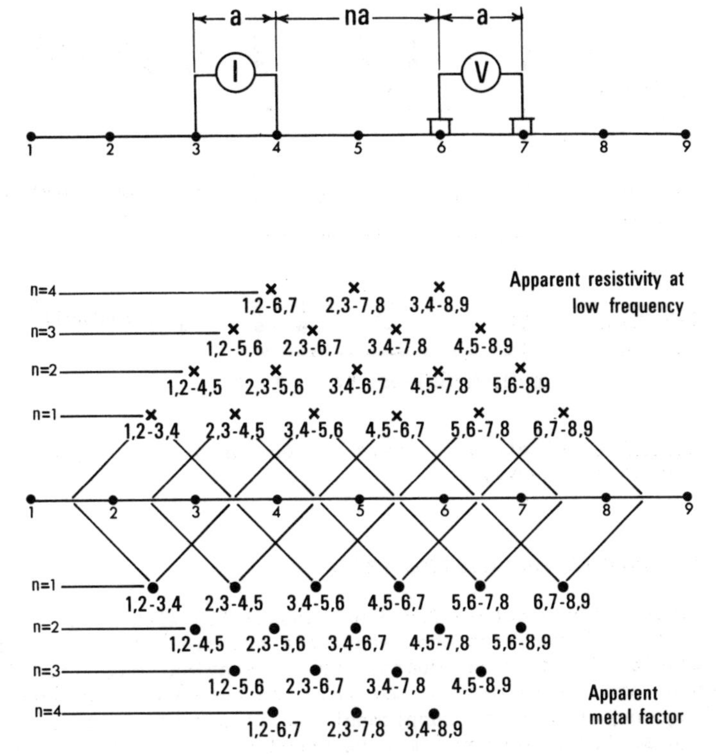

Figure 4.8. Method of positioning point for plotting pseudo-sections in frequency domain method. First and second pairs of numbers beside a point refer to the surface positions of current electrodes and voltage electrodes respectively, at time the reading, to be plotted at the point, was taken. A pseudo-section does not give a true picture of the subsurface distribution of metal factors, resistivities or chargeabilities, a problem which is magnified when trying to tie together, or simply compare, two field surveys in which different values of a have been used.

TYPICAL ELECTRODE CONFIGURATIONS

The most commonly used figurations are - Wenner, three electrode or pole-dipole, dipole-dipole, and gradient or the Schlumberger. These arrays are shown in Figure 4.9.

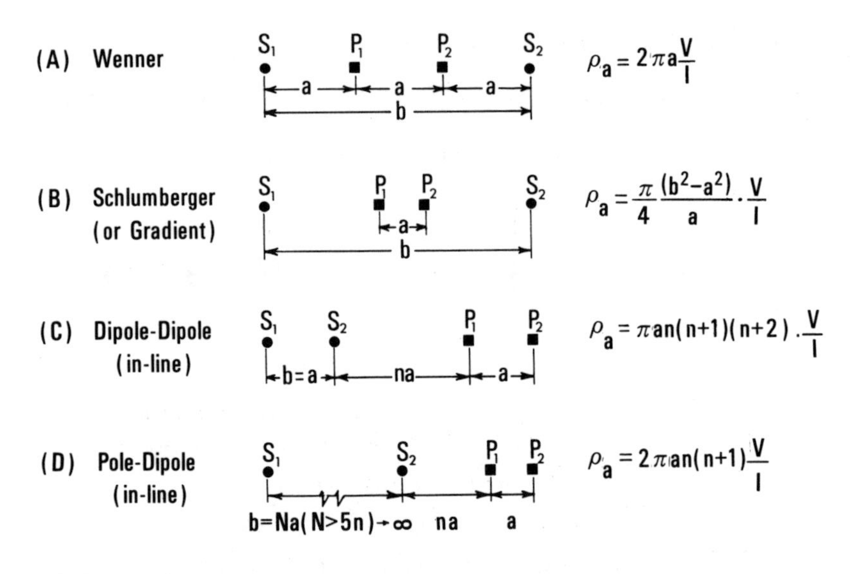

Figure 4.9 Electrode arrays often used for Induced Polarization surveys.

MAGNETIC INDUCED POLARIZATION (MIP)

Since every electric field has associated with it a magnetic field, it follows that the decay in a transient electric induced polarization field is accompanied by a decaying magnetic field. Attempts are being made to use this, in both the time domain and the frequency domain, to supplement the information obtained with the regular I.P., now sometimes referred to as E(lectric) I.P. to differentiate it from M.I.P. data. Both the theoretical and practical aspects of M.I.P. are in the early stages of development and it may be some years before the method is generally accepted as a viable field exploration technique; however, it already appears that if there is an overlying conducting layer the E.I.P. field due to a subsurface I.P. region is considerably reduced relative to the M.I.P. field and that if there is no subsurface chargeability contrast there will be no M.I.P. response, even if the electrical conductivity varies significantly.

SUGGESTIONS FOR FURTHER READING

Edwards, L.S., 1977. A modified pseudosection for resistivity and IP, Geophysics, 42, 1020-1036.

Hallof, P.G., 1974. The IP phase measurement and inductive coupling, Geophysics, 39, 650-665.

Howland-Rose, A.W., Linford, G., Pitcher, D.H., and Seigel, H.O., 1980. Some recent magnetic induced polarization developments - Part I: Theory, Geophysics, 45, 37 - 53.

Pelton, W.H., Ward, S.H., Hallof, P.G., Sill, W.R. and Nelson, P.H., 1978. Mineral discrimination and removal of inductive coupling with multifrequency IP, Geophysics, 43, 588-609.

Seigel, H.O., 1974. The magnetic induced polarization method, Geophysics, 39, 321-339.

Sumner, J.S., 1976. Principles of induced polarization for geophysical interpretation, Elsevier, New York.

Wait, J.R. ed. 1959. Overvoltage research and geophysical applications, Pergamon, New York.

Zonge, K.L. and Wynne, J.C., 1975. Recent advances in complex resistivity measurements, Geophysics, 40, 851-864.

5

Gravity methods

GENERAL IDEAS

From laboratory measurements we know that G, the universal constant of gravitation, is $6.673 \times 10^{-8} \pm 0.003$ cm^3 gm^{-1} sec^{-1}.

It is known that to a first approximation the earth is spherical with a radius R so that since

$$F = -GMm/R^2 \tag{5.1}$$

is the force exerted on a mass m at the surface of the earth, of mass M, then the force F exerted on a unit mass at the surface of the earth is given by

$$g = -GM/R^2 \tag{5.2}$$

where g is called the acceleration due to gravity and the negative sign indicates an inward directed force or acceleration towards the centre of the earth.

g can readily be measured to a high accuracy so that the mass of the earth can be found; M is 5.977×10^{27} gm.

The moment of inertia of the earth about any diameter can also be found; this is done by measuring the response of the earth to a torque exerted by the moon and sun. The Moment of inertia about the polar axis is $0.3307\ MR^2$, a surprising result at first sight since the moment of inertia of a uniform sphere should be $0.4\ MR^2$. The fact that the measured moment of inertia is less than $0.4\ MR^2$ by a significant amount immediately indicates that there is a strong density gradient within the earth, increasing with depth. This evidence is entirely independent of the seismic evidence.

When describing the earth as a sphere to the first approximation, we mean that to an observer far out in space the gravitational equipotential surfaces due to the mass of the earth appear to be spherical, Figure 5.1. As a closer approach to the earth is made it is found that to a second approximation the equipotential surfaces are ellipsoids of revolution about the rotation axis; on even closer approaches it is observed that there are bumps on the ellipsoid and a surface can be fitted to these bumps to any degree of accuracy desired provided sufficient accurate measurements are available.

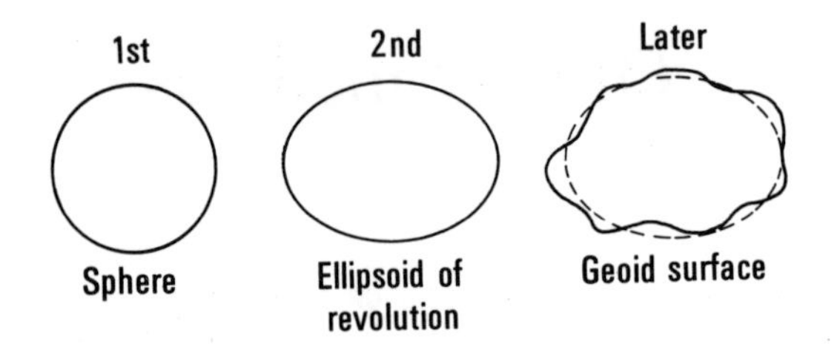

Figure 5.1. Shape of the earth's sea level equipotential surfaces as degree of surface fit increases.

The surface when fitted to mean sea level data is called the geoid. The geoid is therefore the surface which has the same gravitational potential all over the earth as that at mean sea level far out in the oceans. Only rarely does the geoid depart from the ellipsoid of revolution by more than a few meters. The lowest point is in the Indian Ocean off Srilanka where the geoid is 93 meters below the ellipsoid, and the highest point is off Papua - New Guinea where the geoid is 76 meters above the ellipsoid.

Although these elevation differences are quite large the rate of rise and fall is very small so that in exploration work, where the survey area is usually measured in square kilometers, the differences between the geoid and ellipsoid are ignored and all gravity readings and elevations are referred to the ellipsoid of revolution, which is often called the "reference ellipsoid".

From thousands of measurements of g all over the earth we can express the value of g on the reference ellipsoid by the following international formula

$$g_\theta = 978.03185 \, [1 + A \sin^2\theta + B \sin^4\theta] \qquad (5.3)$$

where A = 0.005278895, B = 0.000023462, θ is the latitude, g_θ is expressed in gals (after Galileo) and 1 gal is equal to an acceleration 1 cm sec^{-2}. More accurate formulae can be devised for particular regions of the earth, some including longitude terms.

To compute g_θ from the above formula we have to know the value of the latitude (and longitude if a triaxial formula is used).

Longitude ϕ, is measured in terms of the angle east or west of the plane through the Greenwich meridian.

Geocentric latitude. θ_C is measured by the angle between the line from the centre of the earth to a point on the earth's surface and the equatorial plane, Figure 5.2.

Astronomic latitude. θ_A is the angle between the local vertical, or normal to the geoid, at a point and the equatorial plane. Because of the earth's eccentricity, the geocentric and astronomic latitude are not everywhere the same.

Geodetic or Geographic latitude. θ_B is the angle between the perpendicular to the reference ellipsoid at a point and the equatorial plane.

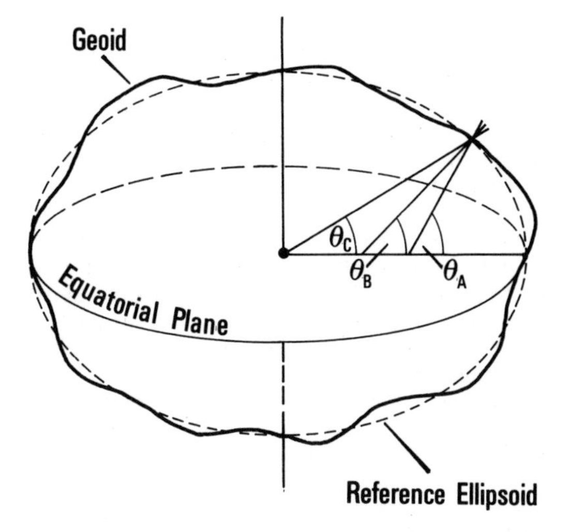

Figure 5.2. Illustrating different ways of measuring latitude. θ_A - astronomic; θ_B - geodetic; θ_C - geocentric.

It is this latter latitude which is used in the formula, although as far as exploration work is concerned the differences between the three are quite small.

THE EXPLORATION PROBLEM

The gravitational acceleration at a point is due simply to the integrated effect of all masses within the range of significance. Much of the gravity interpretation is just a matter of simplifying this integration; as sensitivity and definition requirements increase, so does the complexity of the interpretation increase.

For detailed mapping of the gravitational field on or near the surface, the local inhomogeneities have to be taken into account. Reversing the reasoning, if one knows what the field ought to be if the surface layers were uniform, then by mapping gravity and plotting both theoretical and observed values versus position some idea should be obtained of the size, shape and depth of any inhomogeneities that may be present. This is the basis of gravity prospecting.

If a traverse is made along a section of a semi-infinite plane with no inhomogeneities below it the instruments would give the same reading over any point on it. If, now, a traverse is made over a section which has a spherical iron ore body beneath it, the sphere represents an excess mass since the original sphere of density ρ_1 has been replaced by a sphere of higher density ρ_2. Therefore, a higher value of g will be observed over the body because of the excess mass, Figure 5.3. In practice, it is the density contrast, $(\rho_2 - \rho_1) = \sigma$ which is of interest in addition to the shape and size of the body.

For accurate detailed surface measurements g is not measured absolutely; differences in g are measured by a gravimeter, which is usually an instrument based on a stretching spring principle, Figure 5.3. The gravimeter spring is calibrated in terms of the stretch versus force applied by means of a tilt table or by making observations along a well established calibration line formed by stations where gravity has been accurately measured and the values cover a wide range. Clearly a set of stations from Alaska to Mexico would cover such a range for North America (see equation 5.3).

Because of the extremely high precision required (about 1 part in 10^8) simple extension springs are replaced by much more complex arrangements, but the principle of a strain being used to reflect a stress is unchanged.

Anomaly Due to a General Mass Distribution

In Figure 5.4 the attraction at the point P on the surface due to an elemental mass excess, dm, of the subsurface body, density contrast σ, is $Gdm/r^2 = G\,dv/r^2$ along PQ. The vertical component at P is therefore given by $G\,\sigma dv \sin\theta/r^2$. Therefore Δg

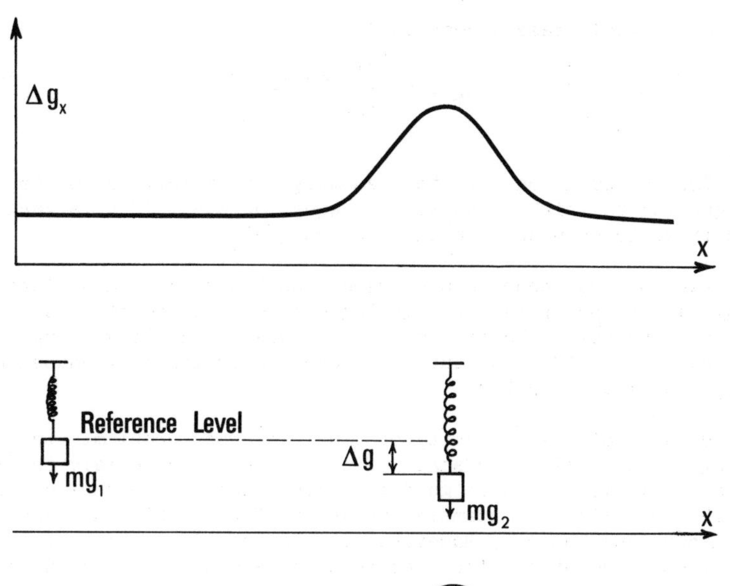

$\rho_1(<\rho_2)$

Figure 5.3. Illustrating the basic principles of origin and measurement of gravity anomalies.

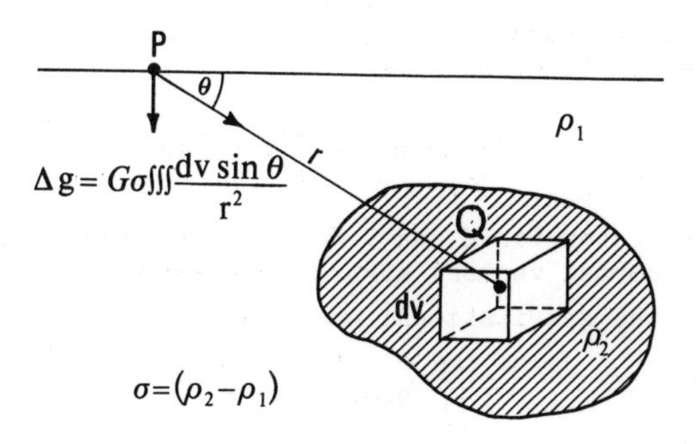

Figure 5.4. Anomaly due to a general mass distribution.

due to the whole mass excess is

$$\Delta g = G\sigma \iiint \frac{dv \sin \theta}{r^2} \tag{5.4}$$

The calculation of the anomaly is simply a matter of integrating the right hand side of equation 5.4. If σ is variable then it would be placed inside the integral.

Various approaches can be used, the two broad classifications being (1) to approximate the geological body to one of ideal shape to give boundary conditions which are relatively simple to integrate, or (2) to integrate over an arbitrarily shaped body using a digital computer.

Because of the ambiguity of the inverse problem, that is trying to determine the shape of a body from the anomaly profile, it is difficult even with good geological control to derive from a gravity profile an unambiguous answer for the size and shape of the anomalous mass. However, it is possible to obtain an unambiguous solution for the total excess mass, which is useful information in estimating ore reserves.

Total Anomalous Mass

Let M be the total anomalous mass made up of excess masses m_1, m_2 ...... m_n, Figure 5.5. Consider the hemisphere enclosing the excess masses. The total normal gravitational flux across the closed surface of the hemisphere is $4\pi GM$ of which half will cross the hemispherical surface and half will cross the circular plane of the ground surface.

By Gauss`s theorem, which states that the volume integral of the divergence (masses) is equal to the total normal amount of flux across the surface, we have

$$\iiint \bar{v}.\bar{n}ds = \iiint \nabla.\bar{v}dv \tag{5.5}$$

where $\bar{v}$ is any vector field and $\bar{v}.\bar{n}ds$ is the normal component ($\bar{n}$ is the unit normal vector) across an element ds of the surface, from which for the gravitational case

$$\iint \Delta g.ds = 2\pi GM \tag{5.6}$$

where Δg is the vertical gravity anomaly measured at a point on the surface.

In practice, the products of the observed average Δg over a small area ds is summed for the whole area so that the total

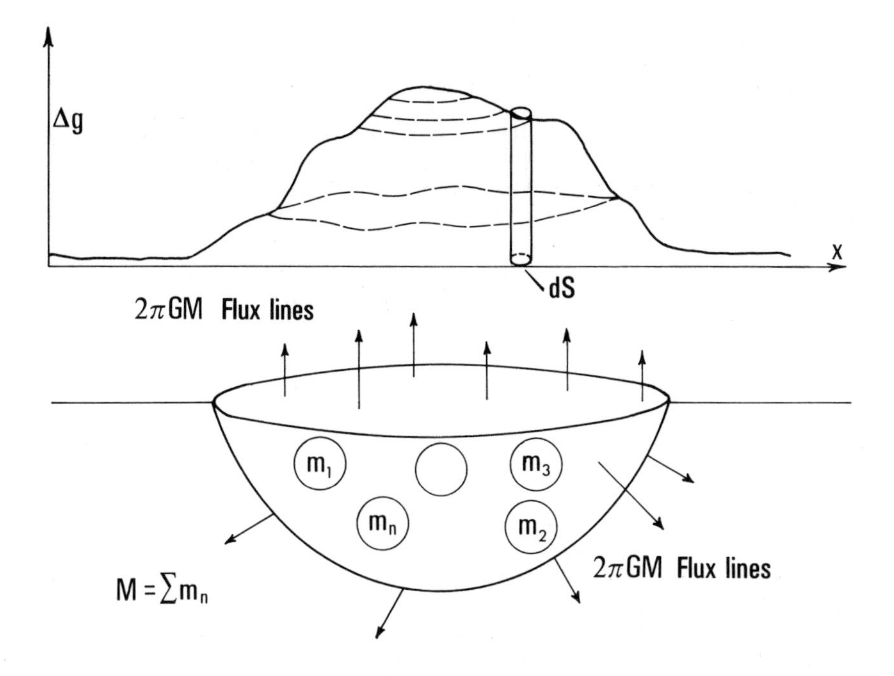

Figure 5.5. Use of Gauss´s theorem to estimate total anomalous mass.

anomalous mass is given by

$$M = \frac{1}{2\pi G} \Sigma \Delta g.ds \qquad (5.7)$$

If the body has a density of ρ_2 and the country rock has a density of ρ_1 then the actual mass M_A is

$$M_A = \frac{1}{\rho_2 - \rho_1} \frac{1}{2\pi G} \Sigma \Delta g.ds \qquad (5.8)$$

It is important to note that the anomaly must tend to zero outside the region of interest otherwise the boundary condition required by Gauss`s theorem is not fulfilled; that is, some of the sources would be outside the area of integration.

CORRECTIONS TO THE FIELD DATA

Since it is very rare to make gravity measurements in a region of constant elevation, it is clear that the effects of varying elevation and topographic masses must be considered.

Free Air Correction

Suppose that a field measurement of g_h is made at the top of a plateau of height h above mean sea level (i.e. h is the height above the geoid or the reference ellipsoid). The distance from the centre of the earth to the point of measurement is now (R+h) where R is the radius of the earth. If g_θ is the value of gravitational attraction on the reference surface at latitude θ then

$$g_\theta = \frac{-GM}{R^2} \quad \text{and} \quad g_h = \frac{-GM}{(R+h)^2} \tag{5.9}$$

Therefore

$$g_\theta = g_h\frac{(R+h)^2}{R^2} = g_h(1+\frac{h}{R})^2 \tag{5.10}$$

or

$$g_\theta = g_h(1+\frac{2h}{R} + \frac{h^2}{R^2}) \tag{5.11}$$

Since h<<R $(h/R)^2$ is ignored. The difference in gravitational acceleration between the two levels due solely to the difference in elevation is therefore $2hg_h/R$ which has to be added to the observed value of g_h to incease it to g_θ, the value of acceleration at mean sea level; unfortunately, this process is frequently termed "reducing" the data to sea level even though when h is +ve the observed value of gravity is increased.

Bouguer Correction

There is also an attraction due to the mass of the elevated region above mean sea level, or above the geoid. To find this attraction it is convenient to first find the attraction at a point on the axis of a disc, since this result will be required for other purposes later, and then increase the radius of the disc to infinity. This attraction will be found by two methods, the first using force fields and the second using elementary potential theory, since the two approaches illustrate two commonly used approaches in more complex problems.

Force field approach

Attraction of elemental mass on a unit mass at Q, Figure 5.6, is $G\rho hxd\phi dx/(x^2+y^2)$ along QR. Therefore attraction of elemental mass along the axis QO is $G\rho hxd\phi dx \cos\theta/(x^2+y^2)$ from which the attraction dA, along QO of the elemental annulus is

$$dA = \int_{\phi=0}^{\phi=2\pi} \frac{G\rho hxdx\cos\theta}{x^2 + z^2}\, d\phi = \frac{2\pi g\rho hx\cos\theta}{(x^2 + z^2)}\, dx \tag{5.12}$$

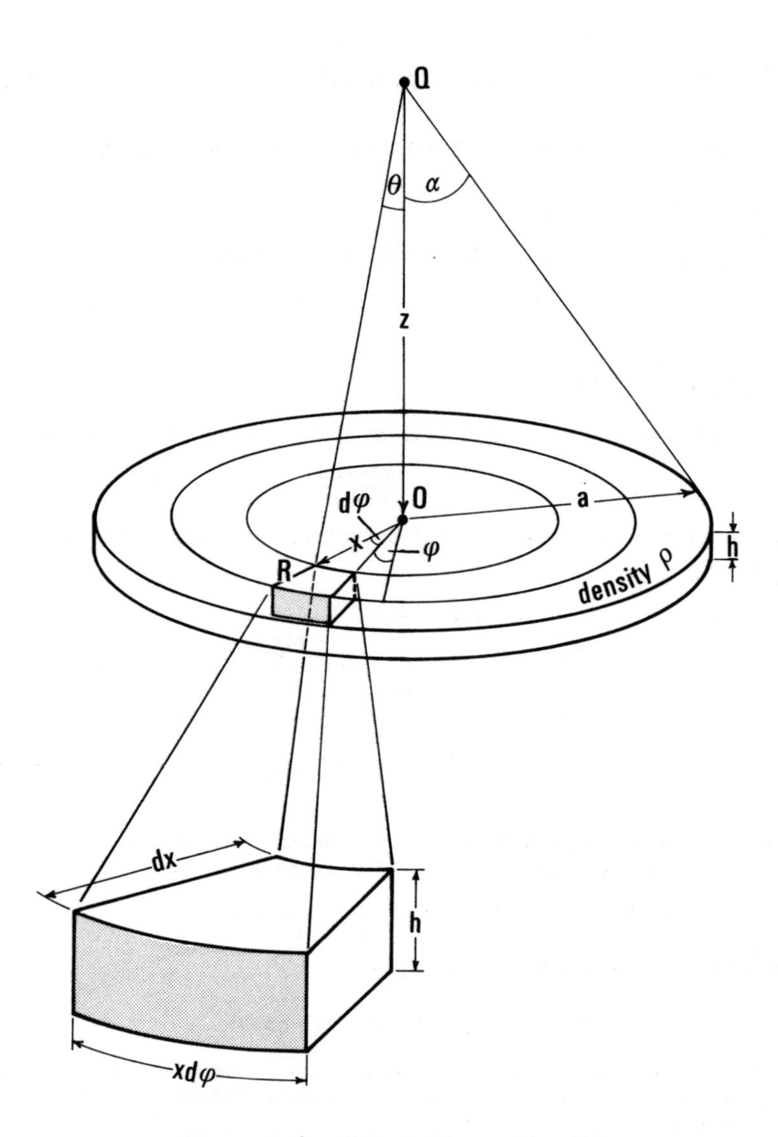

Figure 5.6. Calculating attractions on the axis of a thin circular plate. Mass dm of element = $\rho hx.d\phi.dx$. Values of z and h must be such that both surfaces of the disc subtend essentially the same solid angle at Q.

Since $x = z\tan\theta$ and therefore $dx = z\sec^2\theta d\theta$ substitution in 5.12 and simplification gives

$$dA = 2\pi G\rho h \sin\theta d\theta \tag{5.13}$$

Therefore the total attraction, A, along QO is given by

$$A = \int_{\theta=0}^{\theta=\alpha} 2\pi G\rho h \sin\theta d\theta = 2\pi G\rho h(1-\cos\alpha) \tag{5.14}$$

Potential field approach

The gravitational potential at Q due to the elemental mass is

$$dP = \frac{G\rho h x d\phi dx}{(x^2 + z^2)^{\frac{1}{2}}} \tag{5.15}$$

so that the total gravitational potential at Q due to the complete disc is

$$P = G\rho h \int_{x=0}^{x=a} \int_{\phi=0}^{\phi=2\pi} \frac{x dx}{(x^2 + z^2)^{\frac{1}{2}}} d\phi$$

$$= 2\pi G\rho h \left[(a^2 + z^2)^{\frac{1}{2}} - z\right] \tag{5.16}$$

The force along OQ can be found by taking the gradient of the potential at Q in the direction of z increasing, or

$$F = \frac{\partial P}{\partial z} = 2\pi G\rho h \left[\frac{z}{(a^2 + z^2)^{\frac{1}{2}}} - 1\right] \tag{5.17}$$

Therefore the attraction A along QO is

$$A = -F = 2\pi G\rho h(1-\cos\alpha) \tag{5.18}$$

which is identical with 5.14. Now, if $a \to \infty$ then $\theta \to 90$ or $\pi/2$ and the attraction due to the disc of infinite radius and thickness h is $A = 2\pi G\rho h$.

Therefore the attraction due to an elevated region of infinite extent and of height h above the reference ellipsoid is $2\pi G\rho h$. When the reference surface is the geoid or reference ellipsoid the correction is called the Bouguer correction. However, it is clear that for exploration purposes any other reference surface is satisfactory, the magnitude of the variations being the same. In other words, if a reference surface other than the sea level surface is used, the height axis may be moved up or down leaving the shape of the anomaly curve unchanged.

Technically, if the surface used is not the sea level surface then it should be called the residual anomaly - not the Bouguer anomaly which it has become common practice to do.

Elevation Correction

If the free air and Bouguer corrections are combined, the result is called the elevation correction, equation 5.19.

$$E = \frac{2g_h h}{R} - 2\pi G\rho h = 2h\left[\frac{gh}{R} - \pi G\rho\right] \qquad (5.19)$$

If typical values for g, R, ρ (=2.7 gm cm^{-3}) and G are substituted in 5.19 it is found that

$$E = 0.2 \text{ mgal } m^{-1} \qquad (5.20)$$

where 1 milligal = 1 mgal = 0.001 gal = 0.001 cm s^{-2}.

This has to be added to the observed value of g and applies only to the integrated effects of the masses immediately under the infinite horizontal plane of height h.

It should be noted that for 0.01 mgal accuracy, which many commerical instruments are easily capable of giving, the height h must be measured to better than 5 cm (or 2").

Topographic Correction

In addition to the elevation correction allowance has to be made for topographic features such as valleys and mountains.

A nearby mountain will exert an upward component of force on the gravimeter thereby causing it to read low with respect to what would be read if the plane were infinite, Figure 5.7. Therefore something must be added to correct for this effect.

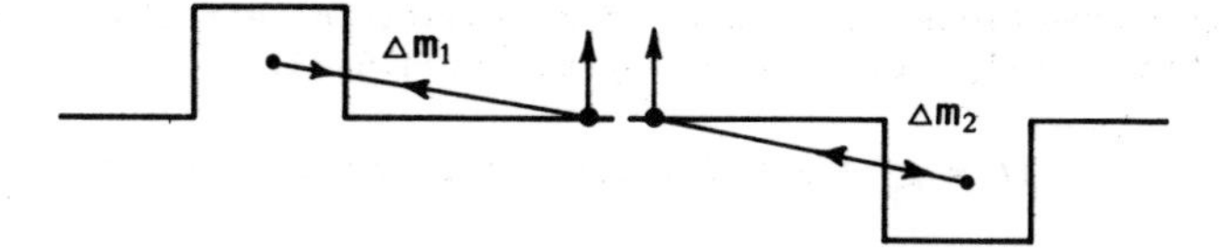

Figure 5.7. Topographic correction. Reasons for correcting for influence of mountains and valleys.

A nearby valley represents a mass deficiency and therefore there is a lower vertical component acting on the gravimeter than there would be if the valley were filled; an alternative approach would be to regard the valley as a region of negative mass.

Therefore the gravity reading will be too low compared with what would be read if the plane were infinite; again, to allow for the valley something must be added to the observed value.

The topographic corrections (T) cannot be made easily unless the topography is sufficiently regular that it can be approximated to a simple geometric shape, for example, an infinitely long mountain range of triangular cross section. In general the corrections are made by computing the attraction of a small raised area after making assumptions about its density and obtaining its volume and approximate centre of gravity from the measured height and base area obtained from maps, then integrating the attractions for a large area in a circular zone about the point at which the observation has been made. The integration can be carried out semi-manually with the aid of zone charts and topographic maps, or with the aid of digital computing programs of varying degrees of sophistication. In principle, the topographic correction is easy to make; in practice it can be quite laborious.

If all corrections have been properly applied to the observed values of g_h then the theoretical gravity g_θ from equation 5.3 should be equal to the right hand side of equation 5.21.

$$g_\theta = g_h + E + T \quad (5.21)$$

In general, it is found that the two sides are not equal and rewriting the equation to read

$$B = g_h + E + T - g_\theta \quad (5.22)$$

gives the Bouguer anomaly, B. In elevated regions B is usually negative indicating a mass deficiency, while in depressed regions it is usually positive indicating a mass excess. The existence of an anomaly may also mean that incorrect densities have been taken in the corrections.

If corrections are made for isostatic effects then the isostatic anomaly, IA is given by equation 5.23.

$$IA = g_h + E + T + \text{isostatic correction} - g_\theta \quad (5.23)$$

The isostatic correction is rarely applied in exploration work and need not concern us here.

ANOMALIES DUE TO SIMPLE BODIES

A Sphere Of Uniform Density

It can easily be shown that the gravitational effect of a sphere of mass M, with either a uniform density or a density distribution which is a function of radius alone, is identical with that due to a point mass M situated at the centre of the sphere.

If the density contrast of the sphere is σ then the force at the point Q on the surface of the earth, Figure 5.8, is GM/R^2 where M is the anomalous mass excess and $= 4\pi a^3\sigma/3$, a is the radius of the sphere and r is the distance from Q to the centre of the sphere. Since most commonly used gravimeters are restricted to detecting only the vertical component of acceleration, the value of the anomalous vertical attraction at Q is given by equation 5.24.

$$\Delta g_v = \frac{GMz}{(x^2 + z^2)^{\frac{3}{2}}} \tag{5.24}$$

The shape of the anomaly curve can readily be found by straightforward substitution for various values of x into equation 5.24, or the equation may be differentiated twice with respect to x to find the maxima, minima and points of inflection. This latter procedure will be followed since it also leads to two simple rules for estimating depths to the centre of the sphere.

From 5.24 we obtain 5.25 and 5.26.

$$\frac{\partial}{\partial x}(\Delta g_v) = \frac{-3GMzx}{(x^2 + z^2)^{\frac{5}{2}}} \tag{5.25}$$

$$\frac{\partial^2}{\partial x^2}(\Delta g_v) = \frac{3GMz(4x^2 - z^2)}{(x^2 + z^2)^{\frac{7}{2}}} \tag{5.26}$$

From 5.25 it can be seen that the first derivative of $\Delta g = 0$ when $x = 0$ and $x = \infty$; inspection of equation 5.26 shows that Δg is a maximum at $x = 0$, a minimum at $x = \infty$ and that the point of inflection occurs at $x = \pm z/2$.

This latter relation is the basis of the first simple depth rule which is that the value of x at the point of inflection on a symmetrical curve like that in Figure 5.8 is half the value of the depth to the centre; in other words, the depth to the centre of the anomalous sphere may be estimated as 2x.

The second simple depth rule is frequently referred to as the "half maximum rule". Let the maximum value of Δg, at $x = 0$, be

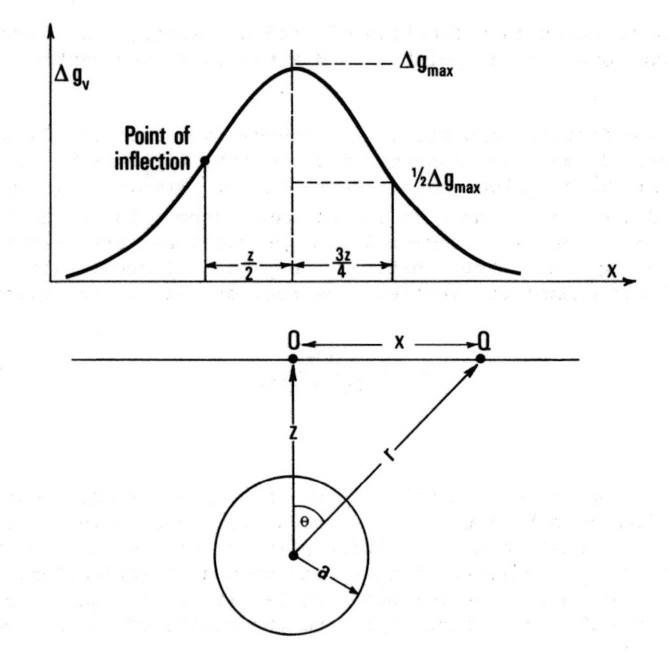

Figure 5.8. Anomaly due to a sphere and illustration of two simple rules for estimating depth to the anomalous mass centre.

Δg_{max} then

$$\Delta g_{max} = \frac{GM}{z^2} \tag{5.27}$$

The value of x when Δg is half the maximum value is given by solving equation 5.28.

$$\tfrac{1}{2}\Delta g_{max} = \tfrac{1}{2}\frac{GM}{z^2} = \frac{GMz}{(x^2 + z^2)^{\frac{3}{2}}} \tag{5.28}$$

From this we can obtain a sixth power equation in z

$$(x^2 + z^2)^3 - 4z^6 = 0 \tag{5.29}$$

which can be factorized into 5.30.

$$(x^2 + z^2 - 4^{\frac{1}{3}}z^2)\ [(x^2 + z^2)^2 + 4^{\frac{1}{3}}z^2(x^2 + z^2) + 4^{\frac{1}{3}}z^4] = 0 \tag{5.30}$$

since both x and z are positive the second term cannot be zero so the first term must equal zero. The first term is a quadratic in z which can be solved to show that $x = 0.766z$. Therefore, with a symmetrical curve such as that shown in Figure 5.8, the depth to the centre of the sphere may be estimated as approximately 4/3 of the distance between the maximum and half maximum gravity anomaly values.

It is important to note that to use either rule it has to be assumed that the symmetrical anomaly is due to a spherically symmetric mass. Symmetrical curves can be caused by other shapes of bodies; it may be easier to prove that a symmetrical anomaly is not due to a spherical mass, by applying both rules and demonstrating that different values of z are obtained.

If vertical gravity gradients are required equation 5.24 can be differentiated with respect to z. Similarly, if the horizontal component of gravity is required, to compare with results from some older types of gravity instruments, then the horizontal component, $GM \sin\theta/r^2$, is used.

<u>Attraction at a Point on the Axis of a Vertical Cylinder</u>

Using the earlier results for a thin disc, equation 5.18, and using dz as the thickness of the disc, we can integrate over a finite length to obtain the attraction due to a cylinder of finite length. Referring to Figure 5.9 the attraction, dA, along the axis due to an elemental volume of the cylinder is given by

$$dA = \frac{G\sigma rzd\phi dzdr}{(r^2+z^2)^{\frac{3}{2}}} \qquad (5.31)$$

so that the attraction due to the complete cylinder is

$$\Delta g = G\int_{\phi=0}^{2\pi}\int_{z=h_1}^{h_2}\int_{r=0}^{a} \frac{\sigma rzd\phi dzdr}{(r^2+z^2)^{\frac{3}{2}}} \qquad (5.32)$$

This is essentially the equation given in 5.4 with the parameters expressed in suitable form for the cylindrical case. If σ is uniform it can go on the outside of the integral signs.

Integrating first with respect to ϕ

$$\Delta g = 2\pi G\sigma \int_{z=h_1}^{h_2}\int_{r=0}^{a} \frac{rzdzdr}{(r^2+z^2)^{\frac{3}{2}}}$$

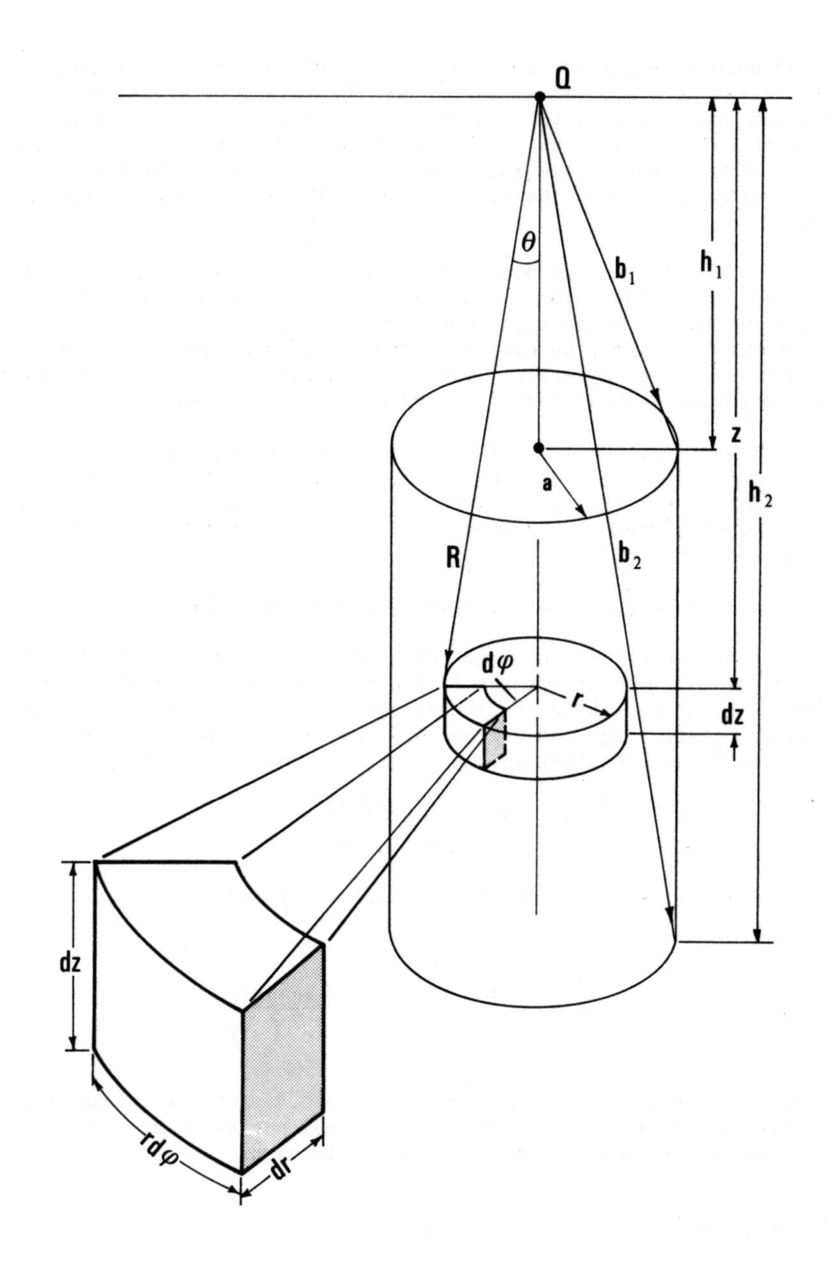

Figure 5.9. Attraction along the axis of a finite cylinder of density contrast σ.

Integrating next with respect to z

$$\Delta g = -2\pi G\sigma \int_{r=0}^{a} r\left[\frac{1}{(r^2+h_1^2)^{\frac{1}{2}}} - \frac{1}{(r^2+h_2^2)^{\frac{1}{2}}}\right] dr$$

Finally, integrating with respect to r

$$\Delta g = 2\pi G\sigma \{[(a^2+h_1^2)^{\frac{1}{2}} - h_1] - [(a^2+h_2^2)^{\frac{1}{2}} - h_1]\}$$

$$= 2\pi G\sigma[(b_1 - h_1) - (b_2 - h_2)] \tag{5.33}$$

If the cylinder is very long then $h_2 \to b_2$ and the equation may be simplified to

$$\Delta g = 2\pi G\sigma(b_1 - h_1) \tag{5.34}$$

which is the result for a semi-infinite cylinder.

Horizontal Sheet of Mass

Let the thickness of a thin sheet, Fig. 5.10, be k and the density contrast σ; the equivalent surface density is therefore $k\sigma$. The contribution to the gravity anomaly of the surface element dS is

$$\frac{Gk\sigma\sin\phi dS}{r^2} = Gk\sigma d\omega \tag{5.35}$$

where $d\omega$ is the solid angle subtended at P by dS.

The total gravity anomaly due to the whole surface is therefore given by

$$\Delta g = Gk\sigma\Omega \tag{5.36}$$

where Ω is the solid angle subtended at P by the sheet mass.

This result can be used to determine the gravity anomaly at points off the axis of regular shape cylinders.

In Figure 5.11 the solid angle subtended at the point P by the elemental disc is $\pi a^2 \cos\theta/R^2 = \pi a^2 z/R^2$.

Therefore the attraction perpendicular to the surface is

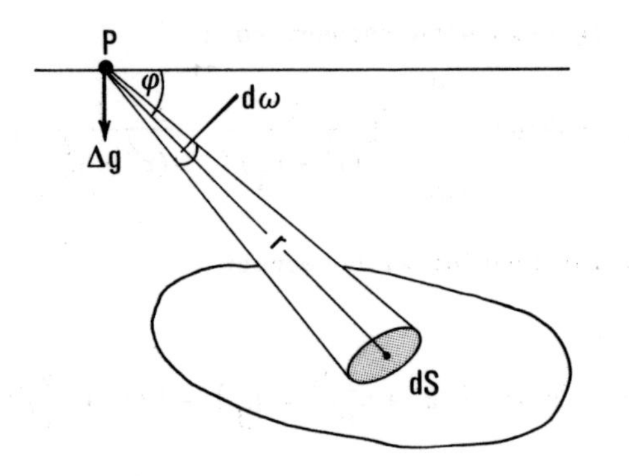

Figure 5.10. Calculation of gravity anomaly using sheet mass and solid angle approach.

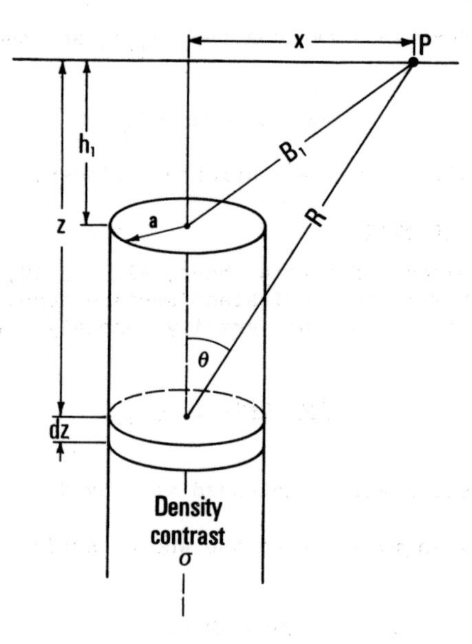

Figure 5.11. Calculation of anomaly due to vertical cylinder at points off the axis.

given by

$$dA = \frac{G\sigma\pi a^2 z}{R^3}\, dz = \frac{G\sigma\pi a^2 z dz}{(x^2 + z^2)^{\frac{3}{2}}} \qquad (5.37)$$

For the attraction due to an infinitely long cylinder, 5.37 may be integrated from $z = h_1$ to ∞ to give

$$\Delta g = \pi G \sigma a^2 \int_{z=h_1}^{z=\infty} \frac{z\,dz}{(x^2 + z^2)^{\frac{3}{2}}} = \frac{\pi G \sigma a^2}{(x^2 + h_1^2)^{\frac{1}{2}}} \qquad (5.38)$$

For a finite cylinder we can take the difference between two semi-infinite cylinders to give

$$\Delta g = \pi G \sigma a^2 \left[\frac{1}{(x^2 + h_1^2)^{\frac{1}{2}}} - \frac{1}{(x^2 + h_2^2)^{\frac{1}{2}}}\right]$$

$$= \pi G \sigma a^2 \left[\frac{1}{B_1} - \frac{1}{B_2}\right] \qquad (5.39)$$

There is an implicit assumption in the above development which requires that for reasonable results B>>a; this follows from the basic assumption that every unit area of the sheet mass subtends the same solid angle at the point being considered. In practice, reasonable results are obtained when B is more than a couple of diameters of the cylinder. Equation 5.39 is therefore essentially an approximation and is used because an exact solution cannot be obtained for off axis points. An alternative approximation, known as the Bowie approximation has been used but it is more complicated to apply and gives less accurate results than the solid angle approach.

Attraction at a Point Due to an Infinitely Long Horizontal Cylinder

The cylinder has a radius a and is buried with its axis horizontal and at a depth z. If the density contrast is σ then

$$F = \text{grad}\ \Phi = \frac{\partial \Phi}{\partial z} = \frac{-2\pi G a^2 \sigma z}{(x^2 + z^2)} \qquad (5.40)$$

where x is the distance from the axis along a line perpendicular to that axis.

THE EOTVOS CORRECTION FOR A MOVING PLATFORM

It is often convenient to measure gravitational fields from a moving platform such as a ship. This introduces motional accelerations which, since a gravimeter cannot differentiate between gravitational and motional accelerations, must be removed from the data. In addition to the obvious accelerations due to pitching, rolling and yawing, there is an additional component due to the steady motion of the platform along the geoid.

Let v be the velocity of the platform at P with components v_n in the north-south direction and v_e in the east-west direction, Figure 5.12. That is

$$v^2 = v_n^2 + v_e^2 \tag{5.41}$$

If the angular velocity of the earth is ω and the platform is at latitude θ, the complete component of east-west velocity is $\omega R\cos\theta + v_e$. The centrifugal acceleration at latitude θ, perpendicular to the axis of rotation, is

$$a_1 = \frac{(\omega R\cos\theta + v_e)^2}{R\cos\theta} \tag{5.42}$$

The centrifugal acceleration a_2 at the same latitude along a radius vector is

$$a_2 = v_n^2/R \tag{5.43}$$

Finally, let a_3 be the centrifugal acceleration due to the earth`s rotation already taken into account in the measurement at a stationary platform, and incorporated in equation 5.3 then

$$a_3 = \frac{(\omega R\cos\theta)^2}{R\cos\theta} = \omega^2 R\cos\theta \tag{5.44}$$

The total extra acceleration, A, along a radius vector through P due to the motion of the platform is given by

$$A = a_1 \cos\theta + a_2 - a_3 \cos\theta \tag{5.45}$$

so that substitution of equations 5.42, through 5.44 in equation 5.45 and subsequent simplification leads to

$$A = \Delta g_c = 2\omega v_e \cos\theta + \frac{v^2}{R} \tag{5.46}$$

where Δg_c is the acceleration correction applied to the observed gravity data. Substitution of typical values for the variables in equation 5.46 gives the results shown in Table 5.1 and shows that the second term is generally insignificant compared with the first term except near the poles. However, both terms are large compared with the easily obtained precision of 0.01 mgal for land based instruments. For a 1 mgal accuracy the east-west component of the velocity should be known to better than 250cosθ m hr^{-1}; at latitude 60° this is 125 m hr^{-1} a fairly high order of accuracy for a ship even if satellite navigation aids are used.

Because the additional short period accelerations due to pitching, yawing and rolling are large they are difficult to reduce to insignificance; the errors in the short period corrections together with those in steady velocity acceleration mean that the accuracy of sea going gravimeter surveys is rarely

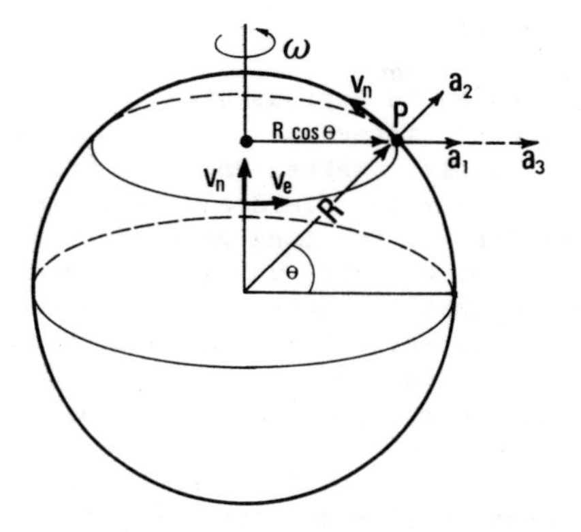

Figure 5.12. Origin of additional acceleration when gravity measurements are made on a moving platform, requiring the application of Eotvos corrections.

Table 5.1

using $\omega = 7.2\text{x}10^{-5}$ radians/second, $R = 6.371\text{x}10^{8}$ cm

$v_e = 3.5\text{x}10^{2}$ cm $s^{-1} = v_n$;

$v = 5\text{x}10^{2}$ cm s^{-1} ($\simeq$ 18km hr^{-1} $\simeq$ 10 knots)

gives $\Delta g_c = 0.05 \cos\theta + 4\text{x}10^{-4}$ gals

θ	$2\omega v_e \cos\theta$ (mgal)	V^2/R (mgal)	g_θ (mgal)
0	50	0.4	50.4
30	43	0.4	43.4
60	25	0.4	25.4
75	13	0.4	13.4
90	0	0.4	0.4

better then 10 mgal. Although the data therefore cannot be used for exploration purposes, they are useful in regional surveys pertaining to oceanic crustal and upper mantle studies. If accurate gravity data are required in oceanic regions, such as in the offshore search for oil, a special type of sea bottom gravimeter is required. This type of gravimeter is essentially a standard land gravimeter, modified by placing it in a pressure proof container, with levelling, zeroing and readout being remotely controlled from the surface.

SECOND VERTICAL DERIVATIVE

Data can be treated in various ways to enchance features that may be of particular interest. One such technique, which is fairly simple to apply, makes use of the second vertical derivative of the anomaly field and is essentially just one of many ways of filtering out the long wavelength variations in order to see the short wavelength variations more easily. It is important to realize that the method, and indeed any method of filtering, does not add any information that is not present in the original data; its use is simply to make some of the gravitational field variations readily apparent to the eye. An approximate value of the second vertical derivative can be obtained from the two orthogonal second horizontal derivatives of the vertical component of the gravitational field.

It can readily be shown that the gravitational field obeys the Laplace equation 5.47.

$$\nabla.\bar{g} = \nabla^2 S = 0 \qquad (5.47)$$

This is itself a scalar field and its gradient $\nabla(\nabla.g)$, may be taken and shown to be zero, so that

$$\bar{i}\frac{\partial^2 g}{\partial x^2} + \bar{j}\frac{\partial^2 g}{\partial y^2} + \bar{k}\frac{\partial^2 g}{\partial z^2} = 0 \qquad (5.48)$$

The field values are in fact not the gravitational field itself but the variation, Δg, in the vertical gravity field so that our final relation, expressed in terms of magnitudes of the orthogonal components, is

$$\frac{\partial^2(\Delta g)}{\partial z^2} = -\left[\frac{\partial^2(\Delta g)}{\partial x^2} + \frac{\partial^2(\Delta g)}{\partial y^2}\right] \qquad (5.49)$$

The right hand side closely represents the curvature, or the inverse of the radius of curvature, of the anomaly surface in two mutually perpendicular horizontal directions. The negative sign indicates that the left hand side is positive when the anomaly surface is concave down, that is where there is a local maximum in the anomaly. For a plane the radius of curvature is infinite so that the right hand side, and therefore the left hand side which is the second vertical derivative, is zero. Regional variations in the field approximate a plane; it is therefore clear that the curvature of anomaly surfaces is much greater for local effects than for regional effects and the second vertical derivative is simply a convenient way of emphasizing local effects. This can be best seen by using a simple analogy.

Consider a simple straight line through the origin, and add to this a sinusoidal function, equation 5.50

$$z = ax + b\sin x \tag{5.50}$$

This function is shown in Figure 5.13(A) where the dotted straight line represents z = ax and may be regarded as the regional or long wavelength component. Differentiating twice with respect to x gives

$$\frac{d^2z}{dx^2} = -b\sin x \tag{5.51}$$

Equation 5.51 is plotted in Figure 5.13(B) and it can be seen that by taking the second derivative we have removed the linear component while retaining the form of the sinusoidal component. That is, we have effectively removed the long wavelength portion of little interest while retaining the short wavelength information.

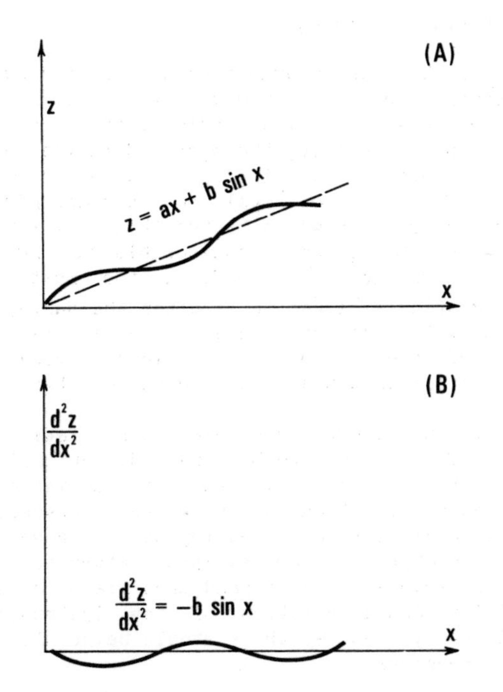

Figure 5.13. Principles of second vertical derivative method; (B) is the second derivative, with respect to x, of (A).

It is worth repeating that the process of taking the second vertical derivative adds no information whatsoever to the original data; it simply presents it in a different way. This is analogous to the various procedures used in presenting seismic data (see Chapter 8).

It is important to note that in taking the second vertical derivative random errors are accentuated; in a dense network an error which is random from station to station may produce large changes in curvature; on the other hand, if stations are only sparsely placed it is not possible to carry out the necessary surface integration with any degree of precision. The second vertical derivative should therefore be used only when there is both a dense distribution of stations and a high precision of measurement at each station.

From equation 5.49 it can be seen that the second vertical derivative can be obtained quite readily from the field data by computing the two orthogonal second horizontal derivatives; the computation can be done by hand or automated.

Outline of Field Procedures

The art of building gravimeters has reached such a stage that it is relatively easy to measure gravity differences of 0.01 mgal, that is differences in the gravitational field can be measured to 1 part in 10^8; in fact, the most modern instruments now claim precisions of 1 μgal (1 part in 10^9). As was shown earlier, for differences of even 0.01 mgal to be meaningful the elevation differences should be known to better than 5 cm (or 0.5 cm for 1 μgal), equation 5.19, and considerable care must be taken with elevation corrections. Therefore, the most time-consuming aspect of gravity exploration is often the surveying required to obtain the elevation differences and, back in the laboratory, the computation of the various corrections; very little time is required to obtain an actual measurement on a typical gravimeter.

It is wise to select a convenient base station somewhere near the centre of an area to be surveyed. Suitable sites for the stations are chosen and their relative elevations determined. Once the station elevations are known the procedure is usually to commence with a gravity meter reading at the base station, make readings at a number of other stations, return to the base station at frequent intervals, which depend upon field circumstances and the stability characteristics of the instrument; generally speaking, in a day`s survey there should be not less than a dozen base station readings.

The base station readings are plotted versus time and the graph obtained is used to correct for possible drift in the instrument and diurnal variations in the gravitational field.

Absolute Values of Gravity

In exploration work absolute values of gravity are not normally required even at the base station. The methods of measuring absolute values of gravity will therefore not be discussed in detail here. Briefly, there are two basic methods, the falling weight method and the pendulum method, the latter being much the most common since the equipment is more easily transportable. However, in both cases numerous corrections are required to allow for departures of the systems from simple boundary conditions demanded by theory, and the precision of measurement is at least an order of magnitude worse than that of difference measurements. That is, even the best method of measuring the absolute value of gravity at any particular point will have a statistical error of about 0.1 mgal.

SUGGESTIONS FOR FURTHER READING

Garland, G.D., 1965. The earth's shape and gravity, Pergamon, London.

Grant, F.S. and West, G.F., 1965. Interpretation theory in applied geophysics, McGraw-Hill, New York.

Nettleton, L.L., 1976. Gravity and Magnetics in oil prospecting, McGraw-Hill, New York.

6

Magnetic methods

REVIEW

We start by recalling some elements of magnetism. In a bar magnet each pole is regarded as being concentrated at a point near, but not at, the end of the magnet. The line joining the poles is called the magnetic axis and the distance between them is called the magnetic length 2ℓ. For a simple bar magnet the magnetic length is usually 5/6 the length of the bar.

The force between two point poles buried in an infinite medium is given by

$$F = \frac{m_1 m_2}{\mu d^2} \tag{6.1}$$

where d is the distance between the point poles of strengths m_1 and m_2, and μ is a constant, called the magnetic permeability, for the medium.

The unit magnetic pole is defined as that pole which when placed 1 cm away from a similar pole repels it with a force of one dyne, both poles being in vacuo (free space) which has a permeability of 1 c.g.s.; for all practical purposes air and a vacuum are the same. Similarly, the permeability of many magnetic rocks is so much larger than that of air or of the host rock that μ is frequently taken as being 1 so that equation 6.1 often is written as $F = m_1 m_2/d^2$. There is a problem here when SI units are used since the value of absolute permeability in free space becomes $4\pi \times 10^{-7}$ Henry/m. By convention north poles are taken to be positive and south poles as negative; with this convention the force is positive between like poles for ferromagnetics (repulsion) and negative for unlike poles (attraction).

Magnetic Force H. The field strength or magnetic intensity H at any point in the magnetic field is the force in dynes which a unit pole would experience if placed at that point. The direction of H is taken to be the direction of the force on a unit north pole. The unit of magnetic force is the oersted (10^{-4} Tesla). The force at a point is one oersted if a pole of unit strength placed at that point experiences a force of one dyne. Therefore a pole of m units of strength in a field of H dynes experiences a force of mH dynes.

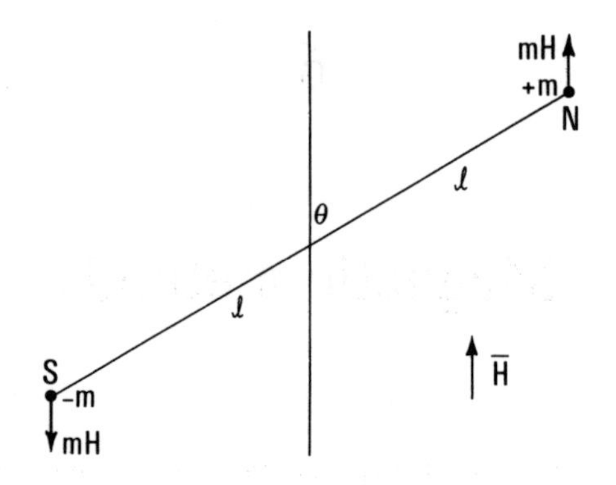

Figure 6.1. Couple exerted on a bar magnet in a uniform field.

In Figure 6.1 a line magnet, represented by NS, has pole strength m and length 2ℓ and is suspended at an angle θ to the direction of a uniform field of H oersted. It can be seen that with the forces on the N and S poles as shown, the magnet experiences a couple C given by

$$C = 2m\ell\ H \sin\theta \qquad (6.2)$$

which tends to swing it into line with the field.

The Couple C depends on three things

(a) the intensity of the magnetic field H which can be regarded as the contribution of the field to the couple
(b) the angle θ which the magnetic axis makes with the field which can be regarded as the contribution of the geometry to the couple
(c) the product $2m\ell$ which is the contribution of the magnet to the moment of the couple and is called the magnetic moment M. That is $C = MH \sin\theta$.

If H = 1 oersted, $\theta = 90^{o}$, then C = M so that M can be defined as the moment of the couple required to hold the magnet at right angles to a uniform field of one oersted. It should be noted that M is a vector quantity since the direction as well as length of the magnetic axis must be specified.

The magnetic moment per unit volume, M/V, where V is the volume, is called the intensity of magnetization, or polarization, I.

The ratio of I to the magnetizing field H is called the

<u>susceptibility</u>, k, of the material so that

$$k = \frac{I}{H} \qquad (6.3)$$

It can readily be shown that the permeability, μ, and susceptibility k are connected by the relation

$$\mu = 1 + 4\pi k \qquad (6.4)$$

The <u>mass susceptibility</u> $\chi = k/\rho$, where ρ is the density and the <u>atomic susceptibility</u> χ_A is the product of χ and the atomic weight of the material.

Intensity due to a Bar Magnet

From Figure 6.2 the forces acting at some arbitrary point B distant r from the mid point of the bar magnet, NS, are $X = m/a_1{}^2$ and $Y = m/a_2{}^2$ in the directions shown. The radial and transverse components are respectively

$$R = X \cos \alpha - Y \cos \beta \qquad (6.5)$$

$$T = X \sin \alpha + Y \cos \beta \qquad (6.6)$$

The following geometric relations will be required in the subsequent substitutions and developments

$$a_1^2 = \ell^2 + r^2 - 2\ell r \cos \theta;$$

$$a_2^2 = \ell^2 + r^2 + 2\ell r \cos \theta$$

$$\sin \alpha = \frac{\ell \sin \theta}{a_1} \; ; \; \sin \beta \frac{\ell \sin \theta}{a_2} \qquad (6.7)$$

$$\cos \alpha = \frac{r^2 - \ell^2 + a_1^2}{2a_1 r} = \frac{r - \ell \cos\theta}{a_1} \; ;$$

$$\cos \beta = \frac{r^2 - \ell^2 + a_2^2}{2a_2 r} = \frac{r + \ell \cos\theta}{a_2}$$

Substitution of the appropriate relations into equations 6.5 and 6.6 gives the following results

$$R = m \left\{ \frac{r - \ell \cos \theta}{(\ell^2 + r^2 - 2\ell r \cos \theta)^{\frac{3}{2}}} - \frac{r + \cos \theta}{(\ell^2 + r^2 + 2\ell r \cos \theta)^{\frac{3}{2}}} \right\}$$

$$T = m\, \ell \sin\theta \{ \frac{(\ell^2 + r^2 + 2\ell r \cos\theta)^{\frac{3}{2}} + (\ell^2 + r^2 - 2\ell r \cos\theta)^{\frac{3}{2}}}{[(\ell^2 + r^2)^2 - 4\ell^2 r^2 \cos^2\theta]^{\frac{3}{2}}}$$

If $\theta = 0$, that is the point B is on the magnetic axis, then appropriate substitution shows that

$$R = \frac{2Mr}{(r^2 - 1^2)^2} \quad \text{and } T = 0 \tag{6.8}$$

when $\theta = 90^o$, that is the point B is on the magnetic equator, appropriate substitution yields

$$R = 0 \text{ and } T = \frac{M}{(r^2 + 1^2)^{\frac{3}{2}}} \tag{6.9}$$

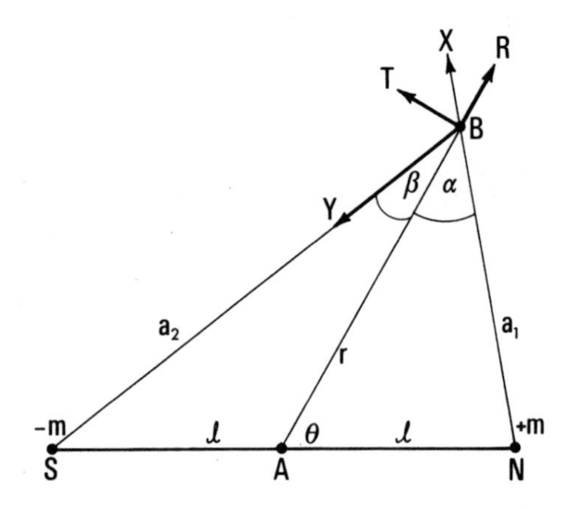

Figure 6.2. Intensity at a point due to a bar magnet.

If $\ell << r$ then for the case when B is on the magnetic axis the magnetic field is all radial and

$$R_r = 2M/r^3 \tag{6.10}$$

whereas when B is on the magnetic equator the field is all transverse, parallel to the magnetic axis, and

$$T_r = M/r^3 \tag{6.11}$$

The field at a point whose distance from the centre of a magnet is much greater than the magnetic length is called a dipole field, although strictly speaking this terminology should apply to the field for all values of r.

THE EARTH AS A DIPOLE

Practical measurements show that the natural magnetic field at the north and south poles is vertical, with little or no horizontal component and approximately equal to 0.6 oersted, while at the equator the field is horizontal with little or no vertical component and approximately equal to 0.3 oersteds. Comparison of these values with the relations in 6.10 or 6.11 shows why the earth's magnetic field is frequently referred to as a dipole field to the first approximation.

That the earth's magnetic field is not a pure dipole field is due to the inhomogeneous magnetic properties of the earth and to the fact that there are a number of time varying changes to the magnetic field. The changes, illustrated in Figure 6.3, in the magnetic field can be classified into two broad groups (a) those with periods greater than a few hundred years which are of internal origin, and (b) those with periods of less than 100 years which are of external origin.

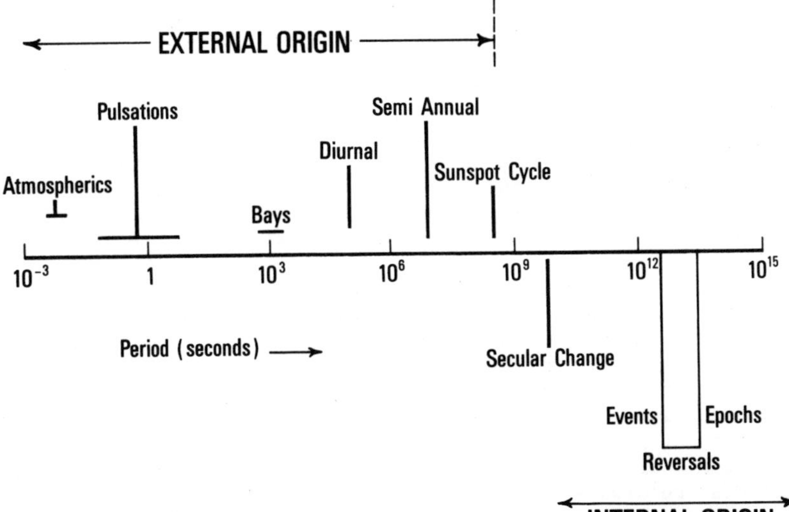

Figure 6.3. Time varying components of the earth's magnetic field.

In category (a) may be listed polar wandering, which is essentially a motion of the dipole field relative to the crust of the earth and which has periods of the order of 10,000 to several million years (the term period being used loosely) and the secular variation, or westward drift, which has a period of one or two thousand years and is essentially a higher order modification of the dipole field.

In category (b), in order or increasing period, we have artificial variations due to nuclear explosions in the atmosphere, micropulsations, individual magnetic storms, bays, diurnal variation, lunar variation, 27 day solar variation, annual variation, and groups of magnetic storms which have an 11.5 or 23 year cycle. In general, in a typical magnetic survey the only variations of concern are bays, diurnal variation, individual magnetic storms and, very rarely, micropulsations. With most magnetic storms little can be done except to cease taking measurements until the disturbances have died down. For the other disturbances it is usually sufficient to take a number of base station readings during a survey; a plot of the variation of the magnetic field with time at the base station is used to correct the field data.

PHYSICS OF MAGNETISM

Since a moving electric charge gives rise to a magnetic, as well as an electric, field it can be expected that a magnetic field will be associated with moving electrons.

The two principal atomic origins of magnetism are (a) the orbital motion of the electrons which give rise to the diamagnetic component and (b) the spin motion of electrons which give rise to the paramagnetic effects; the macroscopic difference between these two effects is explained later.

Diamagnetic properties are exhibited when the orbital electrons are paired or even in number, since the spin moments of the electrons then cancel leaving the orbital moment alone. With odd numbers of electrons, the paramagnetic effects swamp the diamagnetic effect. In both cases the individual moments are usually randomly oriented but application of a magnetic field creates some ordering of the dipoles leaving a net resultant dipole. Since magnetic fields permeate space the effects are always there.

Some elements, in the transition series, have as many as five unpaired electrons in overlapping shells which gives rise to a much stronger magnetic moment; these elements form what are often called "magnetic atoms" or "magnetic ions". Ferromagnetism is exhibited by only the few pure crystalline forms formed by these elements. Among metallic ions that are commonly contained in the natural rocks, only the Ferric (Fe^{+++}), Ferrous (Fe^{++}) and Manganese (Mn^{++}) ions have a natural magnetic moment; cobalt and nickel also have magnetic moments but their abundance in natural rocks is relatively low. None of the other ions have a natural magnetic moment.

Minerals containing the "magnetic ions" noted above generally show ferrimagnetic, antiferromagnetic or paramagnetic properties; minerals containing none of them show only the diamagnetic property.

Diamagnetism

This is exhibited by all materials; $\mu<1$; k is negative and practically independent of the field and temperature; induced magnetism is opposite (antiparallel) to the direction of the inducing field. That is, if the substance is suspended it will move towards a weaker part of the field because like poles are induced opposite one another. Typical naturally occurring diamagnetic materials are quartz and the felspar group.

Paramagnetism

This is exhibited by all materials with unpaired electrons; $\mu>1$; k is positive and practically independent of the field but is temperature dependent obeying the Curie law, $\chi_A \propto 1/T$, where T is the absolute temperature. Induced magnetism is in the same direction (parallel) to the direction of the inducing field. That is if the substance is suspended it will move towards a stronger part of the field because unlike poles are induced opposite one another. Typical paramagnetic minerals are biotite and many pyroxenes.

Special Cases of Paramagnetism

Ferromagnetism. This is exhibited by only a few pure crystalline elements; $\mu>>1$; k is positive and strongly dependent on the field and temperature. At temperatures, T, below the Curie point, Θ, they obey the Curie-Weiss law, $\chi_A = c/(T - \Theta)$ where c is a constant, and a natural alignment of magnetic moments in a very small volume (domain) takes place; at this level a spontaneous macroscopic magnetization is observed. Above the Curie point the susceptibility obeys the Curie law, that is the material behaves as an oridinary paramagnetic. Typical materials are metallic iron, cobalt and nickel.

A block of ferromagnetic or ferrimagnetic material would normally be in a demagnetized state in spite of the spontaneous magnetization in individual domains. This is because the direction of magnetization varies from one domain to another, the total being macroscopically randomly oriented. The application of an external field will tend to order the domains, the ordering increasing with increasing field until saturation magnetization is reached, Figure 6.4.

Antiferromagnetism. Moments of the neighbouring ions are equal in magnitude but opposite in direction (anti-parallel). Therefore there is no net macroscopic magnetic moment.

Ferrimagnetism. This is microscopically similar to antiferromagnetism but macroscopically similar to ferromagnetism since the neighbouring ions are anti-parallel but with one having a moment which is much larger than the anti-parallel moments. The temperature dependence is complicated depending upon the relative intensity of the interaction, but the behaviour is analogous to ferromagnetism.

There are other forms of magnetism which need not concern us at this stage.

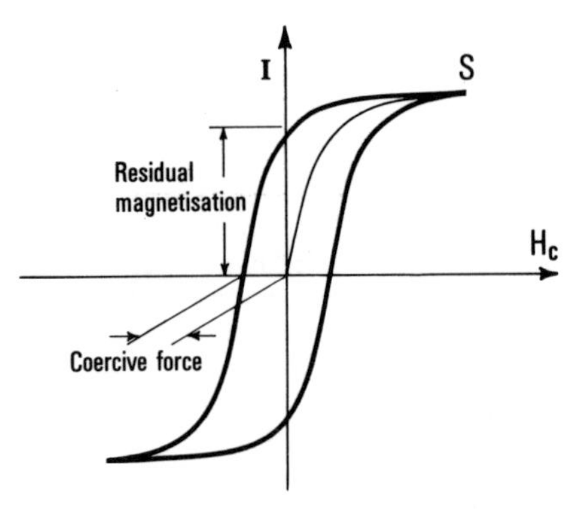

Figure 6.4. Hysteresis curve for a typical ferromagnetic material showing coercive force and residual magnetization.

ELEMENTS OF THE EARTH'S FIELD

In general, the earth's magnetic field is not accurately aligned with geographic north. In Figure 6.5, F is the total field in magnitude and direction with horizontal and vertical components H_0 and Z_0. The angle between the direction of true north and of H is called the declination D, while that between H and F is the inclination I or dip. Since F varies with time all the other dependent quantities also vary with time but the change in D is the most noticeable; this is the reason why most nautical maps indicate the rate of secular variation in D.

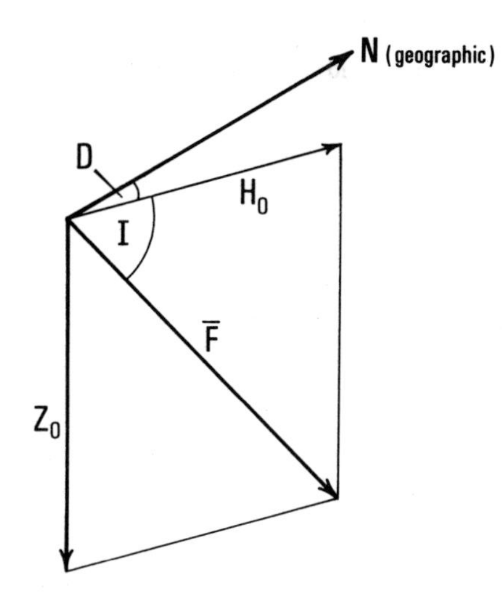

Figure 6.5. Elements of the earth's magnetic field. F - total field; D - declination; I - inclination or dip; H_o and Z_o horizontal and vertical components respectively.

If the crust of the earth was composed of non magnetic rocks there would be no case for the magnetic prospecting method. However, certain rocks exhibit the properties of magnets (either permanent or induced by the earth's field, lightning strikes etc.) and these rock magnets impose their own fields on that of the earth and cause "magnetic anomalies" which can be used to diagnose the form of susceptibility variations in the subsurface and surface.

IMPORTANT GEOCHEMICAL GROUPS

(1) Iron - titanium - oxygen and (2) Iron - sulphur
Magnetite (Fe_3O_4) is ferrimagnetic
Ulvospinel (Fe_2TiO_4) is ferrimagnetic
Ilmenite ($FeTiO_3$) is antiferromagnetic
Hematite (αFe_2O_3) is antiferromagnetic
Maghemite (γFe_2O_3) is ferrimagnetic
Pyrite (FeS_2) is paramagnetic
Troilite (FeS) is antiferromagnetic
Pyrrhotites (FeS_{1+x}, x=0.1 to 0.94) are all ferrimagnetic

Since the intensity of magnetization, I, of ferrimagnetic and ferromagnetic materials depends upon the susceptibility of the rocks or minerals, and since k is field as well as temperature dependent, when typical susceptibilities are given the field strength and temperature at which the susceptibilities were measured must also be given. In practice a unit field is used for reference and the following typical susceptibilities are obtained at normal room temperature

Minerals	Magnetite	- 500,000 x 10^{-6} emu/cc (x 10^{-3} Amps/m)
	Pyrrhotite	- 130,000
Rocks	Basic igneous	- 3000
	Acid Igneous	- 700
	Metamorphic	- 400
	Sedimentary	- 30

The data quoted for rocks are the approximate values for modes in distributions which cover very wide ranges and, in fact, show considerable overlap. This is because for most rocks the magnetic properties can be traced to the presence of magnetite, a most ubiquitous mineral, and occasionally pyrrhotite. The susceptibility of a rock is approximately proportional to the amount of dominant magnetic mineral it contains. For example if a rock contains 0.1 parts of magnetite ($k = 0.5$) then k of the rock is 0.5 X 0.1 = 0.05 emu/cc.

In exploration work the convenient unit of measurement is the gamma (γ) where $1\gamma = 10^{-5}$ oersted (= 1 nT).

REMANENT MAGNETISATIONS

The interpretation of the fields produced by magnetized bodies would be much simpler if the magnetization in crustal rocks consisted entirely of that magnetization which was induced in the material by the present earth`s field. However many rocks possess a permanent or remanent magnetization which may have been acquired very early in their histories and thus in a field which might be very different in orientation and strength from the present earth`s field. The isolation of this remanent magnetization forms the basis of paleomagnetism and it is necessary to give a few working definitions of the types of remanence which can form through natural processes.

Isothermal Remanent Magnetization (IRM) is the induced magnetization obtained by a specimen after applying and removing a magnetic field at a given temperature usually below 400 K. It is time dependent since application of the field for long periods of time increases the magnetization, probably because at the microscopic level thermal agitation helps to order the domains. Conversely if a magnetized specimen is left in field free space for a long period of time, it gradually loses some of its magnetization since in this case continuous thermal agitation will tend to disorder the domains. This time dependent property is called the viscous effect and for this reason it is often called Viscous Remanent Magnetization (VRM). In a rock sample there may be several vector components of IRM due to changes in the direction of the earth`s magnetic field. IRM is fairly easily removed by thermal or AC demagnetization and is therefore called "Soft" or "unstable" magnetization although the magnitude may be large, or "strong".

Thermoremanent Magnetization (TRM) is the magnetization resulting when specimens in a magnetic field are cooled through their Curie point to normal temperatures. The only way to remove the magnetization completely is to reheat the specimen to above the Curie point and cool it in a field free space. It is therefore a very stable or hard magnetization at least as far as a direction of magnetization is concerned, and is the basis of all paleomagnetic results. The magnitude of TRM may also be large. If the specimen is cooled from high temperature, but below the Curie temperature, the result is partial thermoremanent magnetization (PTRM) which is harder than IRM but softer than TRM; the directions of PTRM can sometimes be used in paleomagnetic work.

Chemical or Crystallization Remanet Magnetization (CRM) is caused by the crystallization or chemical formation of ferrimagnetic minerals in the presence of a magnetic field. The growth of ferrimagnetic minerals may take place in igneous rocks by subsolidus exsolution and oxidation processes during slow initial cooling or during a later metamorphic episode. CRM may be formed in sedimentary rocks during the process by which hydrated iron oxides are dewatered to produce the red hematite pigment typical of red beds. CRM is generally a hard magnetization but its usefulness in paleomagnetism is hindered by the difficulty of dating its time of acquistion.

Detrital or Depositional Remanent Magnetization (DRM) is found in sedimentary rocks and is produced when, in deposition processes, particles containing ferrimagnetic minerals, possessing a TRM or CRM by previous processes in their history, are given a preferred orientation due to the tendency of the magnetic axis of the particle to line up with the prevailing direction of the earth`s magnetic field at the time. Because there are numerous disturbing factors such as current directions, shape of the particle etc., the degree of preferred orientation may be small. Therefore the intensity of magnetization in a macroscopic sample may be weak but hard.

Natural Remanent Magnetization (NRM) is the vector sum of all the above remanent magnetizations.

The importance of some of these magnetizations is that a magnetic body may have a mixture of magnetizations. In particular, if a body has received a TRM or PTRM at a time when the earth`s magnetic field is reversed from the present direction, its present NRM might be close to zero since the component of magnetization due to induction in the present field of the earth will tend to cancel the component of TRM. Therefore a body with a high percentage of magnetic material might nevertheless give only a small magnetic anomaly and so be overlooked in a magnetic survey. Such occurrences may be rare but they do exist.

A well planned magnetic survey, particularly one on the ground, should therefore allow for the collection of some oriented hand samples so that laboratory experiments can be made to check the type and degree of magnetizations present.

This being said, all further discussion concerning direction of magnetization will refer to the NRM which might contain PTRM components, although it will tacitly be assumed that in the majority of cases the magnetization is due to induction in the earth`s present field.

DIRECTIONS OF MAGNETIZATION INDUCED BY THE EARTH´S MAIN FIELD

The direction of the induced field can be deduced by considering Figure 6.6.

Remembering that by convention, field lines emanate from a positive N pole and converge on a S pole, some field lines due to the dipole field of the earth are shown. The field induces a magnetization in the body as shown since the dominant characteristic of the material of the body is paramagnetic or stronger (see previous descriptions of types of magnetism). It can readily be seen that at high latitudes the induced field of the body augments the natural field of the earth at the surface; however, at low latitudes the situation is different. This is why when computing the magnetic anomaly due to a particular body, not only should the shape of the body be known, but the direction of the inducing field must be specified.

CALCULATION OF ANOMALY FIELDS

Anomalies due to any size and shape of body can be found by imagining an exploring pole of unit strength on the surface and calculating the magnetic force on it due to the body. To be consistent, a unit positive pole should be used over the entire globe, the field being regarded as positive in the southern hemisphere and negative in the northern hemisphere; then when the induced field in the northern hemisphere augments the natural field, it simply makes the total field more negative, but it would be useful to show the anomalous portion as a positive anomaly. Here we run into a problem of conventions. Due to an earlier convention, it was noted that a force of repulsion between two like poles was positive and by the same convention a S pole, which in the northern hemisphere is what is induced on the upper part of the body, is negative. Therefore to show an anomaly as positive, the imaginary exploring pole of unit strength should be a S-pole

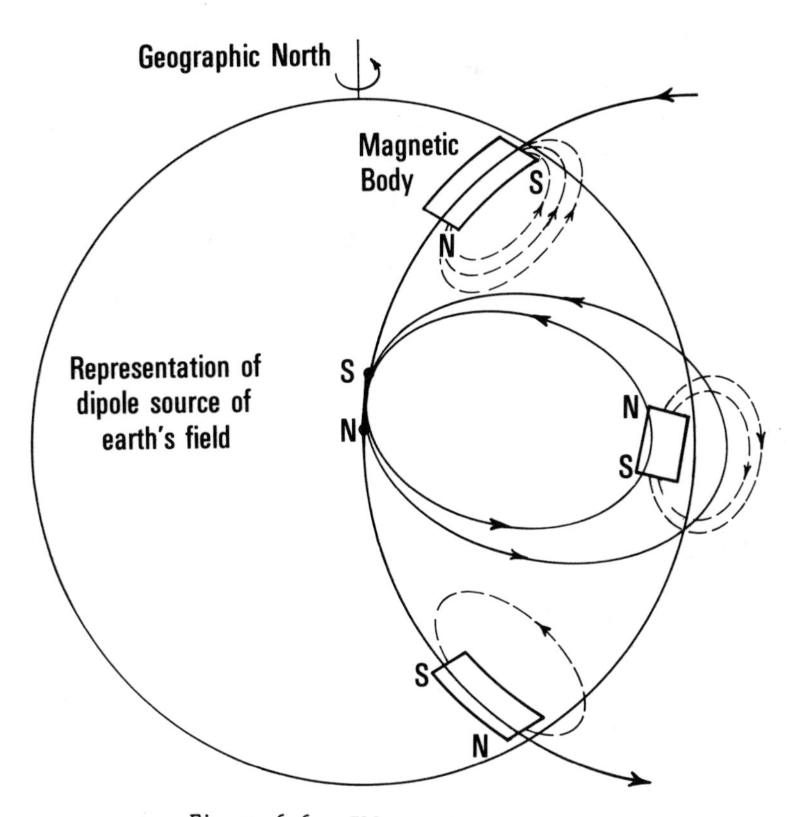

Figure 6.6. Illustrating how induced magnetization in a magnetic body at high latitudes augments the main field, while at low latitudes the main field is opposed.

in the northern hemisphere and a N-pole in the southern hemisphere.

Figure 6.7 shows the vertical (ΔZ), horizontal (ΔH), and total (ΔF) anomaly fields due to a dipole. They are easily calculated by first finding the magnitudes of the fields due to the S-pole then those due to the N-pole, and then adding the two.

These highly specialized cases might seem artificial but if the dipole is extended in a direction perpendicular to the page so that we have a dyke, then anomaly profiles along a line perpendicular to the strike of the dyke will look quite similar to those shown in Figure 6.7, as can be seen from Figure 6.8. In practice, provided the profile is taken three or four dyke widths from the end of the dyke, the approximation to an infinite length is valid.

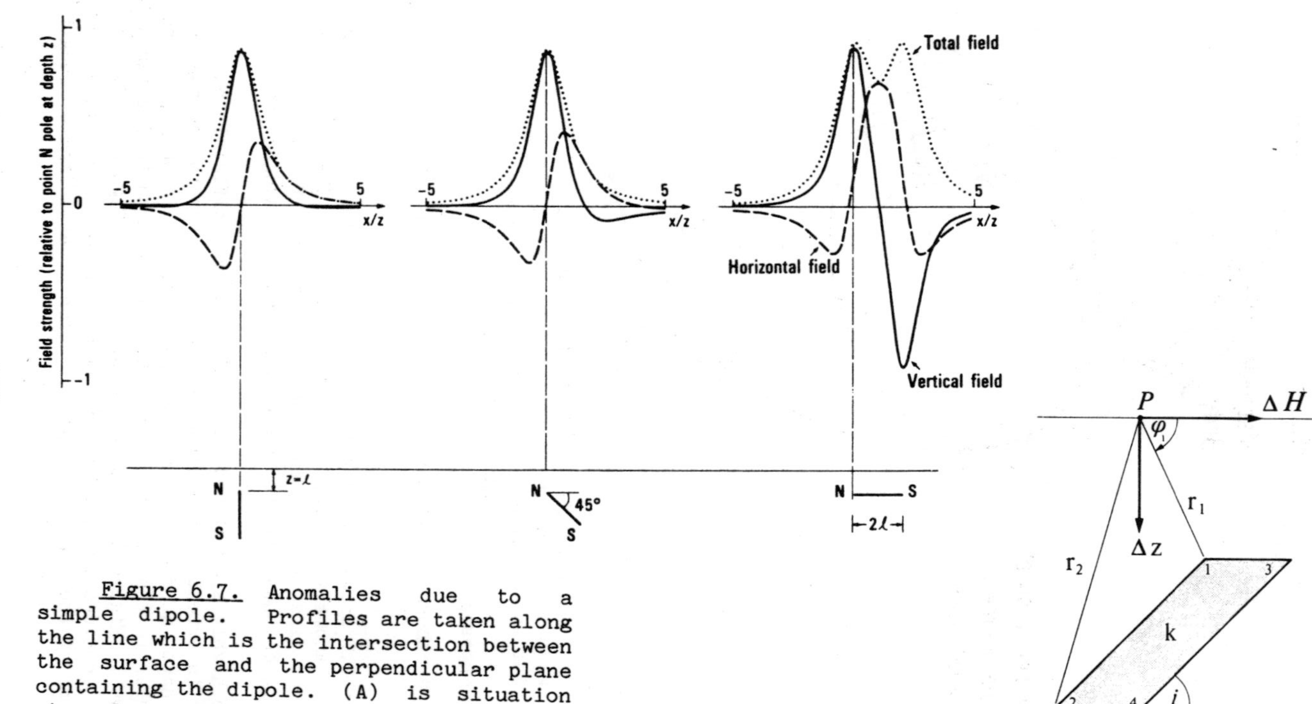

Figure 6.7. Anomalies due to a simple dipole. Profiles are taken along the line which is the intersection between the surface and the perpendicular plane containing the dipole. (A) is situation at polar regions. (C) is situation at equatorial regions.

Figure 6.8. Basis for calculating anomaly curves for infinitely long dykes (after Cook 1957).

The caclulation of anomaly curves for long dykes can be accomplished by building up the procedure used for the simple point pole case. The first step is to calculate the anomaly due to an infinite line of poles of the same sign; the equation is similar to that for the line of mass in the gravity case. Then the line of poles is expanded to a sheet of poles of given surface density; this is similar to the sheet of mass for the gravity case. Next we take another, parallel, sheet of poles of opposite sign and at a different depth, but of the same surface density; for example, the 1-3 and 2-4 surfaces of Figure 6.8. Finally, for the general case of a dipping dyke of rectangular or parallelogram section, we consider two or more pairs of surfaces of poles. After tedious but fairly straightforward calculation, the general equation for the vertical anomaly is

$$\Delta Z = 2k \sin i \, [(H_0 \sin\alpha \, \sin i + Z_0 \cos i) \ln(r_1 r_4 / r_2 r_3) - (H_0 \sin\alpha \, \cos i - Z_0 \sin i)(\phi_1 - \phi_2 - \phi_3 + \phi_4)]$$

where H_0 and Z_0 are the horizontal and vertical components of the inducing field, α is the angle of strike relative to geographic north, i is the angle of dip, k is the susceptibility contrast of the dyke material and r_n and ϕ_n have the meanings shown in Figure 6.8.

TYPES OF MAGNETOMETER

Three types of magnetometer are in current use, each operating on a very different physical principle

(i) Balance type
(ii) Fluxgate
(iii) Nuclear (proton) precession

The first two types are used to measure selected components of the field while the third type is usually used to measure the total field.

Balance type

Here the principle is measuring a torque on an asymmetric magnetic beam and relating it to the force exerted on the beam by the earth's magnetic field, Figure 6.9. Most instruments of this type are adjusted to read zero at a base station selected for convenience and measurements are made of the magnetic field differences relative to the base station value.

The torque can be measured (a) by having a counterweight on the beam and arranging to measure the deflection of a light beam, on a scale that has been calibrated for field strength, when the beam reaches a new equilibrium in a changed magnetic field; (b) by applying a calibrated restoring torque by means of, for example, a fibre rigidly attached to the beam.

Accuracy of these instruments is rarely better than 10γ and in the case of (a) errors of 25γ are common.

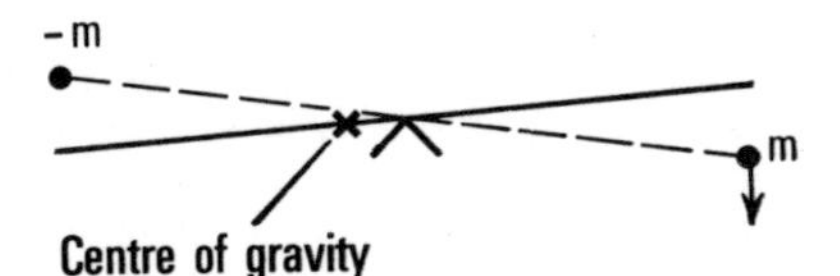

Figure 6.9. Principle of magnetic balance type of magnetometer. Gravitational and magnetic torques are arranged to almost balance each other.

Fluxgate type

The principle of this is best illustrated by taking an instrument with two identical high permeability cores wound with coils connected in series but oppositely wound so as to produce antiparallel magnetic fields along their axes; secondary coils are also wound round the cores, Figure 6.10.

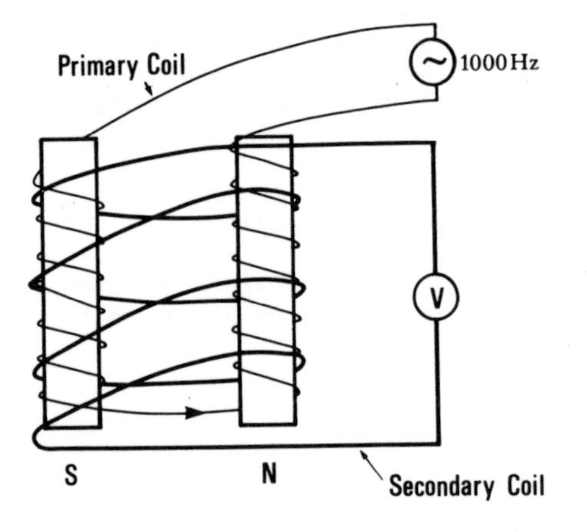

Figure 6.10. Principle of saturated core fluxgate magnetometer.

When a current is passed through the primary coils, they are magnetized in opposite directions so that in one core the geomagnetic field component parallel to the core axis augments the applied field and opposes it in the other.

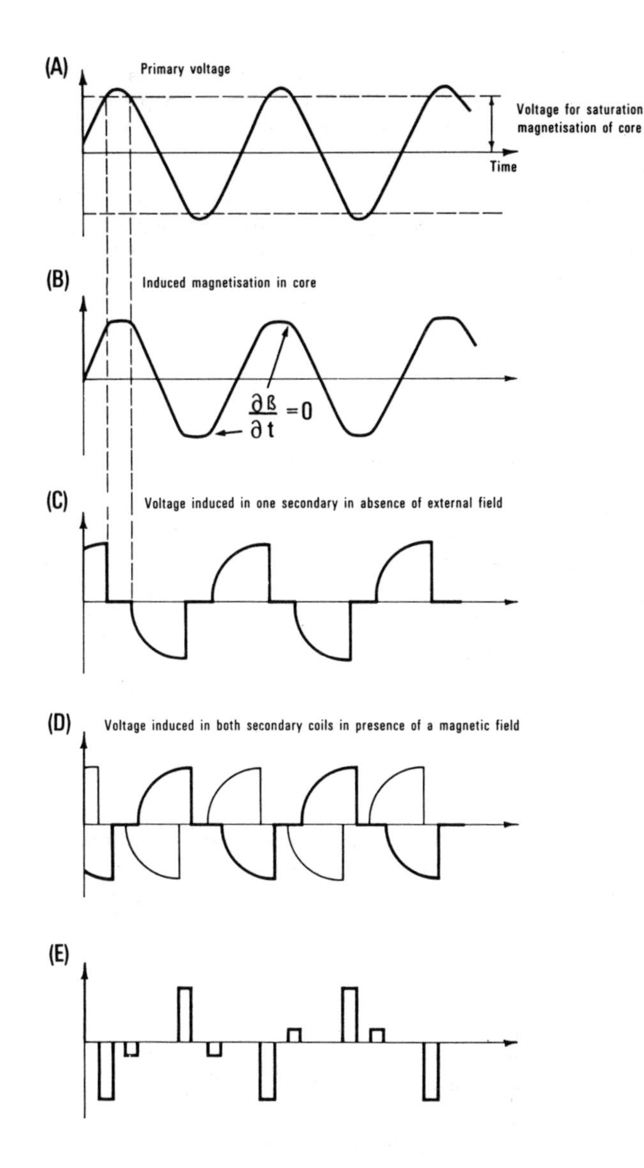

Figure 6.11. Schematic diagram showing primary and secondary voltage relationships for a fluxgate magnetometer (neglecting inductive phase retardation). The peak of the resultant voltage (E) is proportional to the strength of the field parallel to the axis of the coils.

If the current passed through the primary is sufficent to drive the cores to magnetic saturation, that is point S of Figure 6.4 is approached, one core will reach saturation earlier than the other.

The voltage induced in the secondary is the sum due to the two cores; since the induced field is proportional to the rate of change of magnetic flux, which is zero when the core is saturated, one component of the core pair will produce a secondary voltage as shown in 6.11(C). If no external magnetic field is present, the secondary voltage due to the second core would be the mirror image of the first one, these are added vectorially and the total induced field would always be zero.

Now suppose that an external field is present. The fact that it augments the field in one coil means that the coil reaches saturation earlier than it otherwise would; similarly the second coil reaches saturation later than it otherwise would, Figure 6.11(D). Adding these vectorially, we find that there is now a residual series of peaks Figure 6.11(E); the magnitudes of these peaks can be shown to be proportional to the strength of the external field parallel to the core axis.

It can be seen from Figure 6.11(C) and (D) that only one core is really necessary for an instrument provided appropriate electronics are supplied.

Nuclear Type

The explanation of the proton precession magnetometer really lies in the realm of quantum physics but the principle can be readily understood by a simple analogy to a precessing top, Figure 6.12.

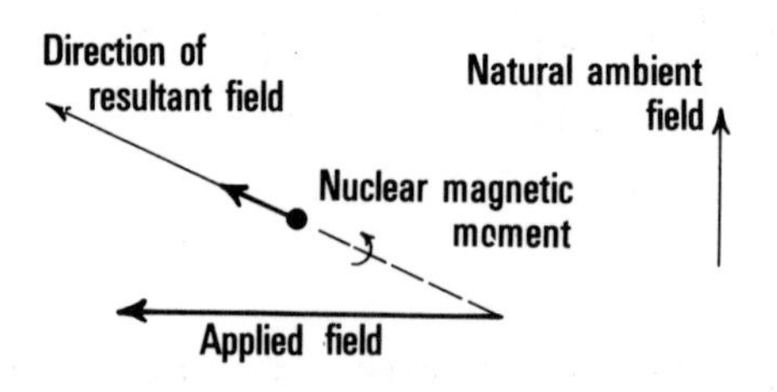

Figure 6.12. Principle of proton precession magnetometer.

Atoms consist of a nucleus of positive charge surrounded by electrons and all nuclei behave as spinning particles having angular momentum and responses like tiny gyroscopes. From quantum mechanics it is known that the angular momentum is quantized; its value can be measured but not predicted.

As might be expected, since the spinning nucleus can be regarded as a moving charged particle, it has a magnetic field associated with it. The moment of this field is also quantized, can be measured but cannot be predicted.

Since the nucleii possess a magnetic moment, they tend to line up in the direction of the ambient magnetic field. If the direction of this field is changed suddenly, the nuclei will try to follow the field. Since they possess angular momentum they cannot do this instantaneously and, acting as spinning tops, they will begin to precess about a mean spin axis which gradually changes direction to line up with the new field direction. It can be shown that the rate of precession is inversely proportional to the angular momentum of the nucleus and directly proportional to the nuclear magnetic moment and to the strength of the new field. That is, for a given nucleus, the frequency of precession depends only on field strength.

Since the precessing nuclei set up a rotating magnetic field which is resolvable into two sinsusoidal quadrature (90° out of phase) components at right angles to each other in space, a coil placed around them will have a voltage induced in it, the frequency of the induced voltage being the precession frequency.

The precession rate, or Larmor frequency, $\omega = \gamma^{*}H$ where γ^{*} is the gyromagnetic ratio (nuclear magnetic moment/nuclear angular momentum) and H is the external field. If the precession frequency is $f_p(=\omega/2\pi)$ then the strength of the external field is given by

$$H = \frac{1}{\gamma^{*}} \cdot 2\pi\ f_p = \frac{2\pi}{\gamma^{*}} \frac{N_p}{T} \tag{6.12}$$

where N_p is the number of precessions in time T.

In practice the coil which produced the large applied field is also used for determining f_p. This means that the applied field current must be switched off so rapidly that the nucleii cannot follow the changing resultant field direction. During the off period the precession frequency is measured. With an appropriate switching arrangement measurements can be made every two or three seconds.

For a group of nuclei precessing in phase, there is obviously some interaction and the precession amplitude is gradually reduced by a number of damping forces, such as other accelerations, thermal agitation and field gradient. The relaxation time varies from about 10^{-5} seconds to a few tens of seconds depending on the nucleus and the amplitude of the signal depends on the number of nuclei per unit volume. Fortunately, water is a nearly ideal candidate for the active material; the relaxation time of hydrogen is a few seconds and the number of nuclei per unit volume is high while oxygen, in common with all nuclei having even atomic mass and even atomic number, has no spin and therefore plays no active part in the experiment. Kerosene is also much used in proton magnetometers.

The precession frequency for hydrogen is 4257.60±0.03 Hz/oersted; that is, in the earth`s field the frequency to be measured is around 2000 Hz. It can also be seen that 1 Hz corresponds to about 25γ or that to obtain 1γ accuracy the frequency must be determined to 0.04 Hz.

The accuracy of frequency measurement increases with the time allowed for its measurement; on the other hand, the signal is gradually damped out, and increased measurement time means a lower rate of repetition of measurement which could be important in airborne surveys. Some form of compromise between these factors is therefore always necessary.

Strong field gradients will also affect the measurements since the nuclei at one end of the detector head will then precess at a different frequency from those at the other end; this means the rate of damping is increased. Field gradients that can be tolerated for given accuracies on a typical instrument are

for 0.5γ accuracy 200γ/m
1.0γ accuracy 400γ/m
2.0γ accuracy 800γ/m

These are not really very large gradients but could be encountered in ground surveys in some mining areas, so that accuracy is sometimes limited by them.

FIELD PROCEDURE

For ground based surveys a convenient base station is selected somewhere near the centre of the area to be surveyed (it is assumed that in dense bush country, lines have already been cut). Base station readings should be taken every 1/2 hr to 1 hr and a plot made of the readings versus time. The curve so obtained is used to correct all station readings for diurnal variation, which can reach 30γ in high latitude areas, and instrument drift. This latter variation is not really significant for good modern instruments. The results are contoured or profiled and interpreted by comparison with type curves.

For airborne surveys the base station is usually a recording magnetometer in, or not too far from, the field area. A typical field pattern flown is shown in Figure 6.13, the flight lines being perpendicular to the suspected strike of the formations with 1/2 to 2 km spacing. The triply flown base line, on which the ground recording base station should be placed, is used to give the diurnal variation correction curve using the data obtained at various times at the flight path intersection points. This method can be used even without a ground base station although it is safer to do both.

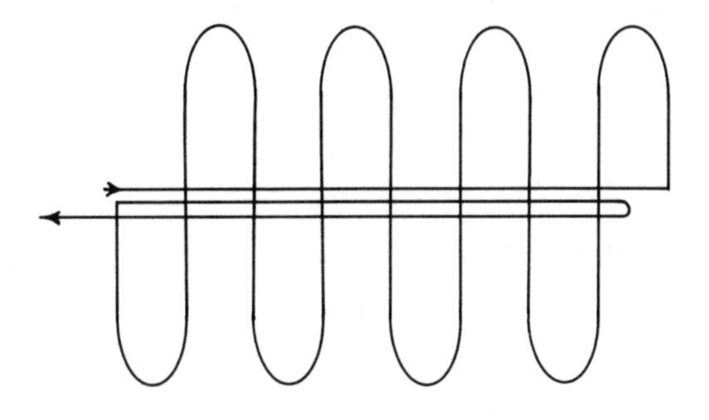

Figure 6.13. Typical flight pattern for an airborne magnetometer survey. Note numerous intersections which can be used to check time variations of the field data.

At the same time as the magnetic data are being recorded the height of the plane and its position are being recorded, the latter by means of photography or radar.

High accuracy of magnetic measurements is not normally required in airborne surveys, especially over mining areas, since the position errors may swamp the magnetic errors. For example, if a typical flight velocity is 200 - 250 km/hr the plane will travel a distance of about 60 m in one second. If a nuclear precession magnetometer is being used with a readout every 2 seconds, the plane has travelled over 100 m between readings. An accuracy of 1γ will therefore be quite unnecessary, although it is fairly easy to obtain with modern instruments.

In both land based and airborne surveys, if the base station readings indicate that a magnetic storm is in progress during the survey, the wisest action is to stop taking readings until the storm is over. However, it is possible that future development of magnetic gradiometer surveys may eliminate this problem. Gradiometer surveys are carried out using two closely spaced ($\simeq$10m), very high sensitivity (accuracy about 0.01γ) magnetometers and measuring the magnetic field gradient directly. This approach has several advantages, but one major disadvantage is that a new body of interpretation procedures have to be built up.

SUGGESTIONS FOR FURTHER READING

Carmichael, C.M., 1964. The magnetization of a rock containing magnetite and hemoilmenite, Geophysics, 29, 87-92.

Cook, K.L., 1957. Quantitative interpretation of vertical magnetic anomalies over veins, Geophysics, 15, 667-686.

Gay, S.P., 1967. Standard curves for interpretation of magnetic anomalies over long tabular bodies, pp. 512-548 in Mining Geophysics Vol. II eds. D.A. Hansen, R.E. MacDougall, G.R. Rogers, J.S. Sumner and S.H. Ward, Society of Exploration Geophysicists, Tulsa, Okla.

Grant, F.S. and West, G.S., 1965. Interpretation theory in applied geophysics, McGraw-Hill, New York.

Nettleton, L.L., 1976. Gravity and magnetics in oil prospecting, McGraw-Hill, New York.

Smellie, D.W., 1967. Elementary approximations in aeromagnetic interpretation, pp. 474-489, in Mining Geophysics Vol. II, eds. D.A. Hansen, R.E. MacDougall, G.R. Rogers, J.S. Summner and S.H. Ward, Society of Exploration Geophysicists, Tulsa, Okla.

Stanley, J.M., 1977. Simplified magnetic interpretation of the geologic contact and thin dike, Geophysics, 42, 1236-1240.

Strangway, D.W., 1970. History of the earth's magnetic field, McGraw-Hill, New York.

7

Electromagnetic methods

BASICS OF WAVE PROPAGATION

If a plane, unattenuated wave of circular frequency $\omega(=2\pi f$ where f is frequency in Hz) is propagating with velocity u, the amplitude at z = 0 at time t is given by

$$A = A_o \cos \omega t \qquad (7.1)$$

Therefore at any position z in the direction of propagation, the amplitude is given by

$$A = A_o \cos \omega(t - \frac{z}{u}) \qquad (7.2)$$

where u is called the phase velocity because it is the velocity we can associate with a fixed value of the phase, i.e. it represents the speed of advance, along the z axis, of a point of zero relative displacement, for example 1/2 A_o or A = 0 in Figure 7.1(A). $\omega(t-z/u)$ is called the phase angle or, for shortness, phase.

The wave equation is often written as

$$A = A_o \cos (\omega t - k_1 z) = A_o \cos 2\pi(\frac{t}{T} - \frac{z}{\lambda}) \qquad (7.3)$$

where $k_1 = \omega/u$ and is called the wave number (or circular wave number), the wave length, λ, is the distance over which $k_1 z$ changes by 2π radians, and T, the period, is the inverse of the frequency, f.

In exponential or complex number notation the wave quation may be written as

$$A = A_o e^{i(\omega t - k_1 z)} \qquad (7.4)$$

where, remembering that

$$e^{i\theta} = \cos\theta + i \sin\theta \qquad (7.5)$$

the meaning of equation 7.4 is that A is the real part of the right hand side.

If the plane wave is, as is usual, undergoing attenuation as it travels along the z axis and its amplitude decreases exponentially, Figure 7.1(B), then we can allow for this by rewriting equation 7.4 as

$$A = (A_o e^{-bz})\ e^{i(\omega t - k_1 z)} = A_o e^{i(\omega t - k_1 z) - bz} \qquad (7.6)$$

or

$$A = A_o e^{i(\omega t - kz)} \qquad (7.7)$$

where $k = k_1 - ib$. k is still called the wave number in this general case, but it is complex with the imaginary part corresponding to attenuation. It is important to note that for an attenuating wave travelling in the positive z direction the real part of k must be positive and the imaginary part must be negative or else the amplitude would grow as z increases. Thus both k_1 and b must be positive.

The quantity ik is often written as γ and called the propagation constant. b is called the attenuation constant, with 1/b being the attenuation distance.

ELECTRO-MAGNETIC WAVES

For electromagnetic (EM) waves γ depends upon the magnetic permeability (μ) the electrical conductivity (σ) and the dielectric constant or relative permativity (ε) of the medium through which the wave is propagating.

Any propagating EM wave contains an electric and a magnetic vector at right angles to each other and forming a plane perpendicular to the direction of propagation, Figure 7.2.

In the electromagnetic method we require two coils, a transmitter (Tx) which may be fixed or portable and a receiver (Rx) which is usually portable. An alternating current is applied to the transmitter which produces an electromagnetic field of the same frequency. Most electromagnetic methods make use of only the magnetic component of the electromagnetic field so generated.

This alternating magnetic field arises and collapses once each cycle so that if there are any conductors (e.g. the Rx) in the vicinity of the transmitter they will be cut by a time varying magnetic flux and therefore have eddy currents induced in them. This concept is analogous to the better known one of a conductor moving through a stationary magnetic field and having a potential difference produced across it. The eddy currents induced in any subsurface conductors will also produce a magnetic field, called the secondary magnetic field, which interferes and combines with the transmitted, or primary, magnetic field to produce a resultant of the two.

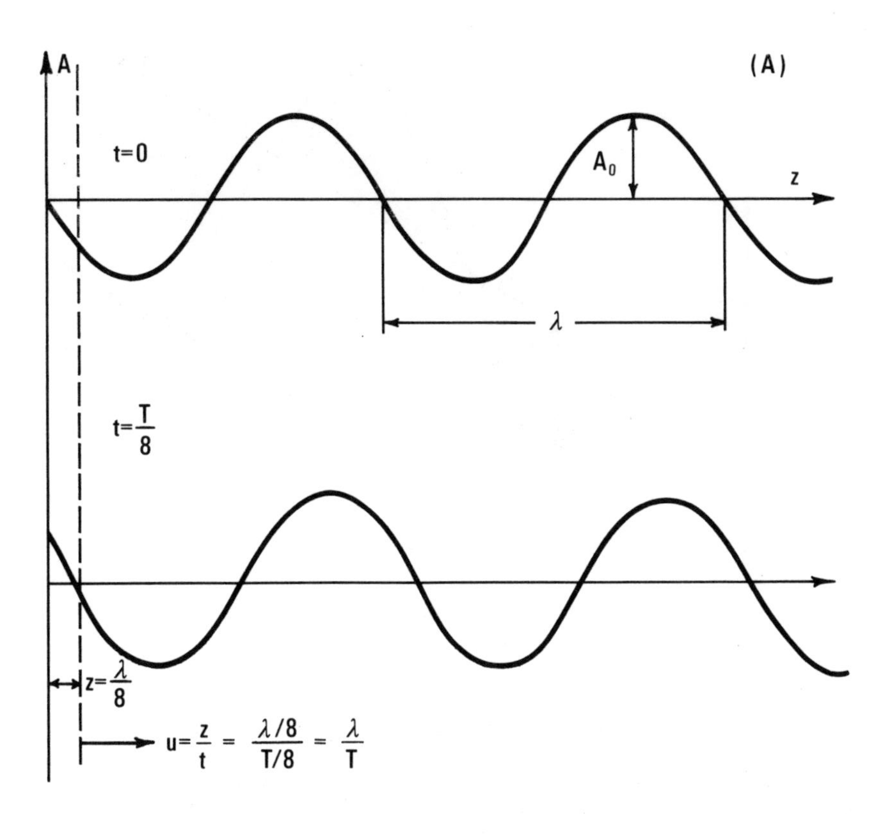

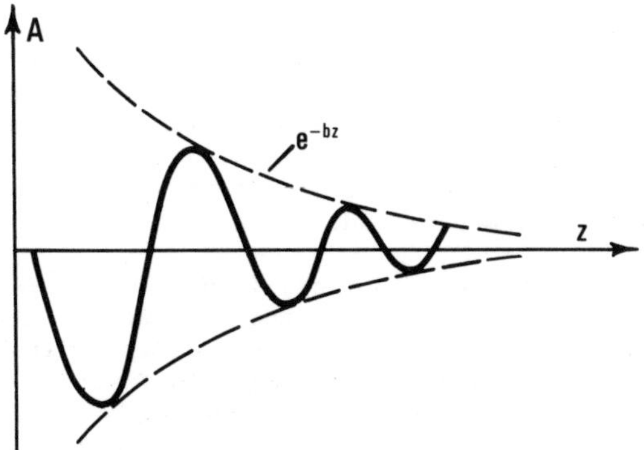

Figure 7.1. Illustrating (A) the meanings of phase velocity u, wavelength λ, period T and amplitude A_0 for a wave of the form $A = A_0 \cos 2\pi (t/T - z/\lambda)$. (B) attenuated wave of the form $A = (A_0 e^{-bz}) \cos 2\pi (t/T - z/\lambda)$.

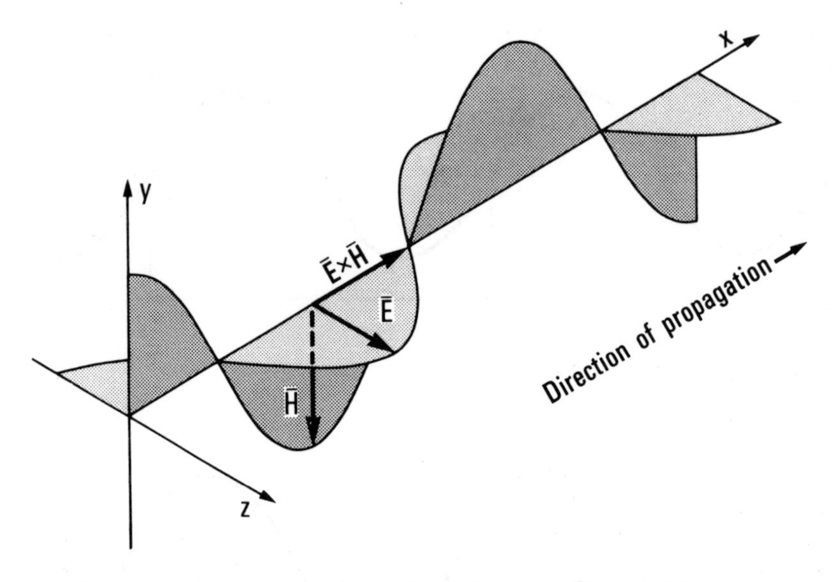

Figure 7.2. Elements of an electromagnetic wave.

BASIS OF THE ELECTROMAGNETIC METHOD

The electromagnetic method depends essentially on the fact that in average ground an applied EM field of a given frequency produces no significant eddy currents, whereas conductors, such as massive sulphide ore bodies, highly conducting ground waters etc., do produce significant eddy currents. The secondary and resultant fields are also oscillating fields and any coil placed in them will have an alternating current induced in it; therefore the receiver really acts as an alternating magnetometer so that techniques which use only the magnetic component of the EM field are really alternating magnetic field methods.

Except in very special cases the frequency, f, should be less than a few thousand Hz. The distance between the transmitter and receiver, both of which are often simple coils, is usually between a few tens to a few hundred meters. From these two factors we find that for a frequency of, say, 3,000 Hz the wavelength λ is given by $c = f\lambda$, where c is the velocity of light and equal to 3×10^5 km/sec, so that λ is about 100 km.

Therefore the distance between the Tx and the Rx is at least two or three orders of magnitude less than the wavelength. This gives an important advantage in the electromagnetic method since it means that in the region of measurement phase retardation, see Figure 7.3, and attenuation are negligible and the effects of propagation can be disregarded. This may be contrasted with the seismic method which depends on the propagation properties of waves. In this sense the electromagnetic method is closer to stationary field methods such as the gravity or magnetic methods than to the seismic method.

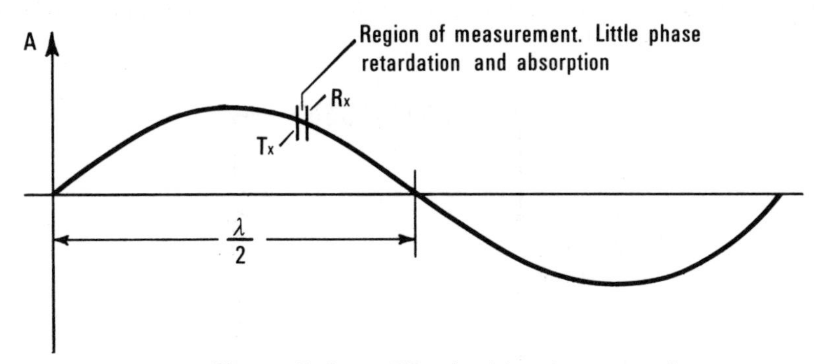

Figure 7.3. Illustrating how normal transmitter (Tx) and receiver (Rx) separation are small compared with wavelength of EM waves of frequences <3,000 Hz, thus allowing effects of propagation to be ignored.

However, the electromagnetic method does depend on phase relationships between the primary and secondary fields when conductors are present; because the currents in the subsurface conductors may be at some arbitrary angle to the transmitter and receiver, and out of phase with the primary magnetic field, the resultant field is, in general, elliptically polarized and various characteristics of the polarization ellipse can be used to make useful deductions about the subsurface conductors.

CIRCUIT THEORY ANALOGUE OF PHASE SHIFT

It may be helpful to recall some of the elements of circuit theory in order to understand the behaviour of the alternating magnetic fields produced at the surface by subsurface bodies. Referring to Figure 7.4 let the alternating voltage be represented by $E = E_o \sin \omega t$. Then the current I in the circuit is

$$I = \frac{E_o}{\sqrt{(\omega L - \frac{1}{\omega C} + R^2}} \sin(\omega t - \alpha) \quad (7.8)$$

where

$$\alpha = \tan^{-1} \frac{(\omega L - \frac{1}{\omega C})}{R}$$

and the negative phase difference (i.e if α is positive) means that I lags behind E

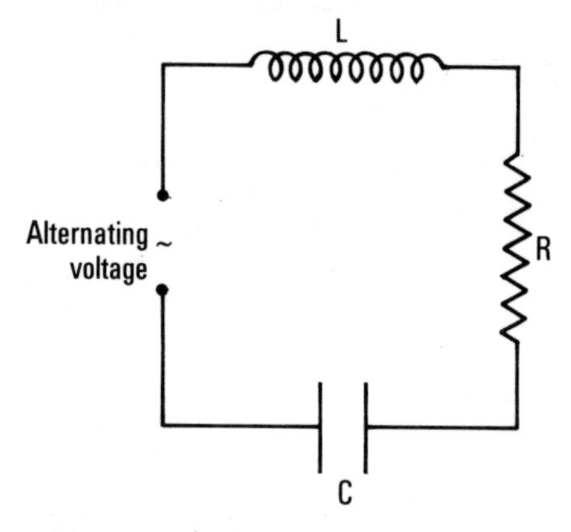

Figure 7.4. Basic electric circuit containing capacitance, C, inductance, L, and resistance R.

If $1/\omega C > \omega L$ then α is negative, the phase difference is positive, and I leads E. If $C \to \infty$, i.e. $(1/C) \to 0$ then

$$\alpha = \tan^{-1} \frac{\omega L}{R} \tag{7.9}$$

Recalling the definition of capacity, that it is the charge required to raise the potential of a body by unity, this means that an infinite voltage would be required for a body with $C = \infty$. Obviously, with a perfect conductor it would not be possible to charge the body at all; in other words a perfect conductor may be regarded as one having an infinite capacity. Therefore in general good conductors can be regarded as possessing a high capacity as well as a low resistance. Even for poorer conductors with high resistance, a high capacity can result because of the size of the body.

If in addition to $C \to \infty$ we have $R \to 0$ then we can write that $\alpha \to \tan^{-1} \infty$ or $\alpha \to \pi/2$ and that for good conductors I lags behind E by a quarter cycle or $\pi/2$. If $R \to \infty$, $\alpha \to \tan^{-1} 0$ or $\alpha \to 0$ and we can state that for bad conductors I is in phase with E.

These generalizations are illustrated in Figure 7.5 where 7.5(B) is a time plot of the primary magnetic field, (frequency $f=\omega/2\pi$), due to the current in the primary conductor, with the general equation

$$P = H_o \sin\omega t \qquad (7.10)$$

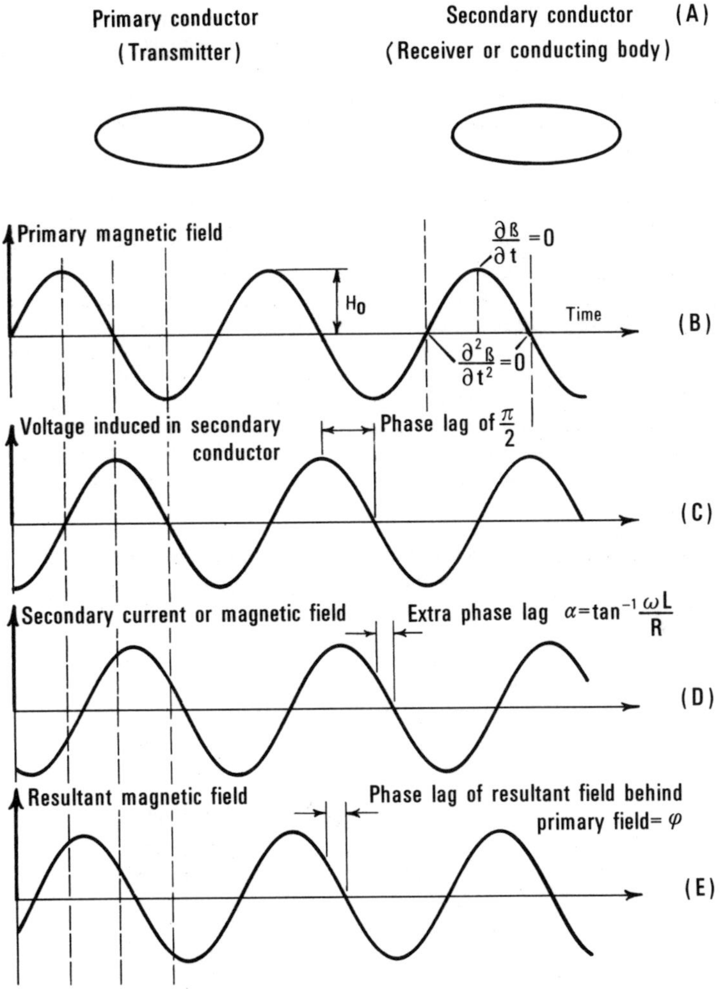

Figure 7.5. Illustrating concept of induced voltages and phase lag between primary, secondary and resultant magnetic fields, which are in phase with the currents, in a conductor.

The voltage induced in a second and perfectly conducting body (i.e. the secondary voltage) lags behind P by $\pi/2$ or 1/4 cycle, but is, of course, of the same frequency. That a $\pi/2$ lag occurs can be seen mathematically by remembering that the induced voltage is the time rate of change of magnetic flux which, at the maxima and minima of the primary field, is zero, Figure 7.5(B); therefore the induced voltage will be zero at a time when the primary magnetic field is at a maximum or minimum, Figure 7.5(C). Similarly, when the second time derivative of magnetic flux of the primary field is zero, the time rate of change of the induced voltage is zero; since the latter value occurs at a time when the primary magnetic field changes from positive to negative or vice versa, the secondary magnetic field will be a maximum or minimum at this time.

Because a conductor usually has a resistance, R, and inductance, L, - the inductance being essentially the tendency of the conductor to oppose a change in the magnetic field linked with it - the current in the conductor will, in general, lag behind the induced voltage by an angle given by equation 7.9. The secondary magnetic field, which is in phase with the secondary current, therefore lags behind the secondary (induced) voltage by the same angle; using these relationships the secondary magnetic field is indicated in Figure 7.5(D). The primary and secondary fields combine in a complex manner and Figure 7.5(E) shows the resultant which is the vector sum of Figure 7.5(B) and (D), the resultant having a phase lag with respect to the primary field, of some angle ϕ.

From equation 7.9 it is clear that changing any one, or a combination, of ω, L, or R can change the phase angle α.

Vector Diagram

These relationships can also be represented in what is known as a vector diagram, shown in Figure 7.6.

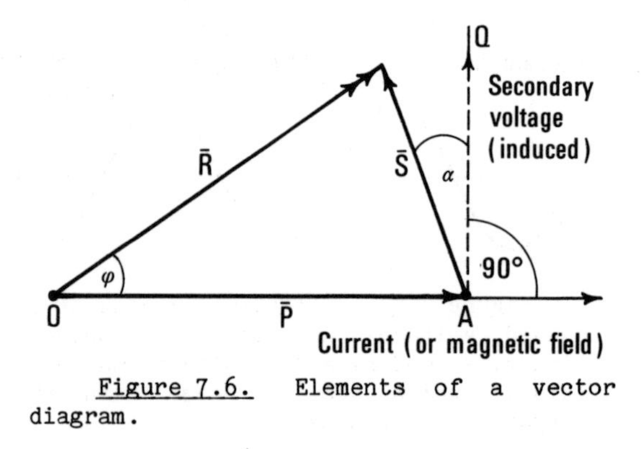

Figure 7.6. Elements of a vector diagram.

Let OA be the amplitude of the primary field, i.e. at the instant in time when the primary field reaches maximum OA will be the vector representing the primary. By convention any vector with an anti-clockwise angle relative to the positive OA direction lags behind OA by the value of the angle. Therefore in a good conductor the induced voltage will lag the primary magnetic field by an angle of 90° which is represented by the vector perpendicular to OA; the secondary current in the conductor, and therefore the secondary magnetic field, will lag a further angle α behind this because of the finite inductance and resistance. This is shown by the vector S. From the parallelogram law we find the resultant R which, from the diagram, lags ϕ behind P.

For good conductors $\alpha \to 90$ and $\phi \to 0$; for poor conductors $\alpha \to 0$ and $\phi \to 0$, and these relations can be used as a rough indication of the conductivity of the subsurface conductors.

POLARIZED WAVES

The vector diagram represents time relations only and it is now necessary to visualise the space relations of these fields.

Figure 7.7(A) is simply a reminder that wave motion can be represented by imagining a rotating arm with the projection of the circumferential point onto the diameter being the apex of the vector. If the arm rotates at constant angular velocity, ω, the point moves with simple harmonic motion and represents the variation of the magnitude of the wave with time.

Plane Polarized Waves

Now consider two vectors $\bar{P}$ and $\bar{S}$ in phase and perpendicular to one another in space, at an instant in time (t_1) when their magnitudes are at a maximum, Figure 7.7(B); their resultant is given by $\bar{R}(t_1)$. At some later instant in time, t_2, they will both have the same magnitudes but less than their maximum; their resultant will be $\bar{R}(t_2)$. Similarly at some later time t_3 the resultant will be $\bar{R}(t_3)$. It is clear that the tip of $\bar{R}(t)$ moves in a straight line that passes through the origin, 0, and the resultant is therefore linearly polarized.

For two similar vectors in phase, but with unequal amplitudes and at an angle β, instead of at 90 , to one another it is again easy to see that the tip of the resultant vector will move on a straight line going through the origin to give a linearly polarized field, Figure 7.7(C).

In general, therefore if the primary and secondary vectors are in phase the resultant will be a linearly polarized wave in a plane at an angle to the primary plane which depends upon the amplitudes of, and the angle between, the $\bar{S}$ and $\bar{P}$ vectors. It can readily be seen that the same comments apply if $\bar{S}$ and $\bar{P}$ are 180°

out of phase.

Circularly and Elliptically Polarized Waves

Now consider the case when $\bar{S}$ and $\bar{P}$ are of the same amplitude are perpendicular to each other but 90° out of phase. This can be easily represented by a circular diagram similar to that of 7.7(A).

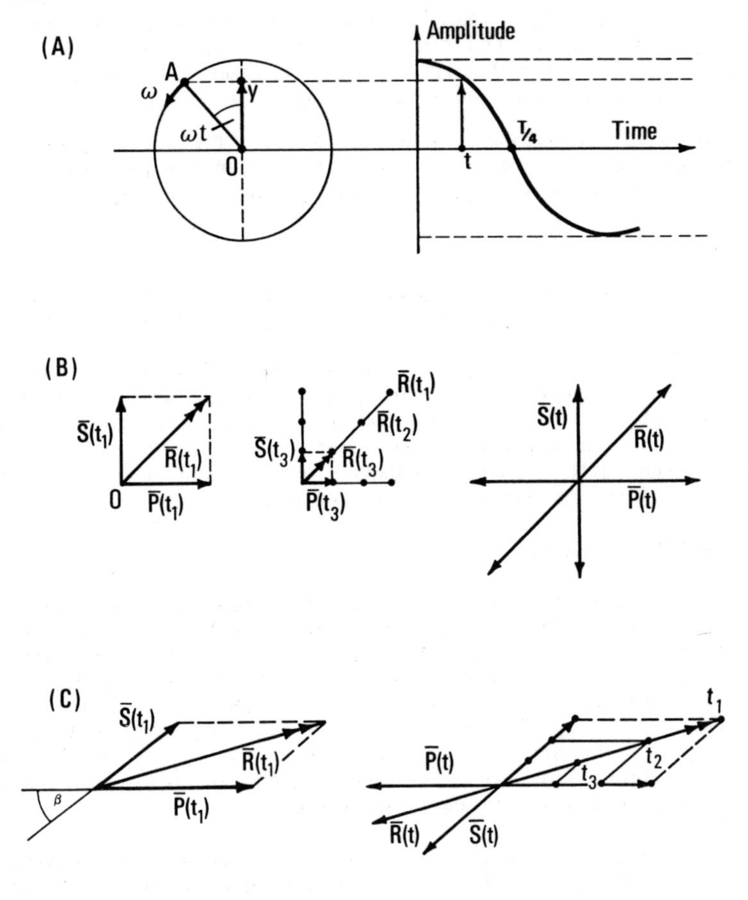

Figure 7.7. (A) Representation of simple harmonic motion. Also indicating how two vectors in phase, (B) of equal magnitude and perpendicular to each other in space, and (C) of unequal magnitude and at some arbitrary angle to each other in space, combine to give a linearly (plane) polarized resultant.

Referring to Figure 7.8(A) if $\bar{S}$ and $\bar{P}$ are out of phase then the tip of $\bar{P}$ will be on the projection of Q on to OP and the tip of $\bar{S}$ will be on the projection of Q on to OS. Therefore the resultant $\bar{R} = \bar{P} + \bar{S}$ but since Q moves on a circle then the amplitude of $\bar{R}$ is constant. That is the tip of $\bar{R}$ moves with Q and the resultant is a circularly polarized magnetic field.

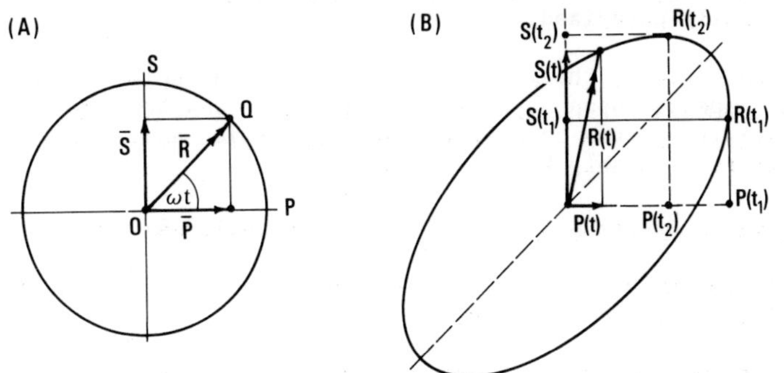

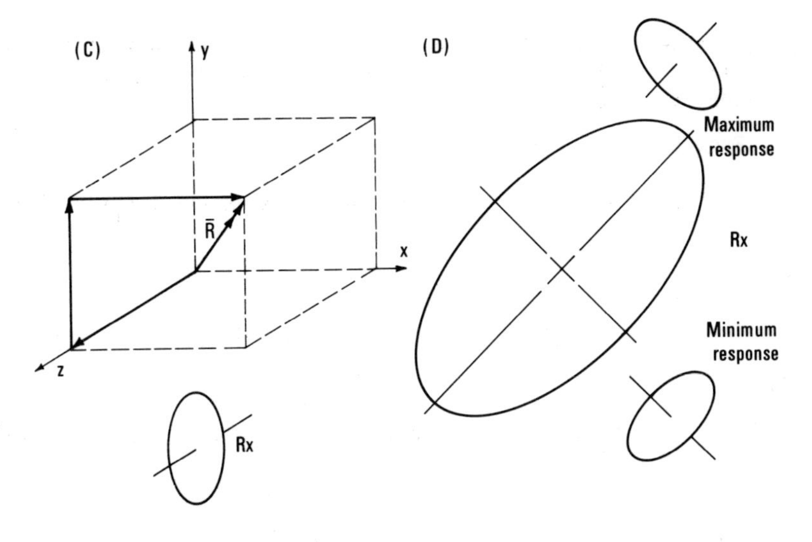

Figure 7.8. Showing how (A) circularly and (B) elliptically polarized fields arise: in (B) the primary and secondary fields need not necessarily be at right angles to each other; the vector notation has been dropped for clarity. (C) and (D) illustrate how response of R is proportional to component of field parallel to coil axis.

Now consider the more general case where $\bar{P}$ and $\bar{S}$ differ by a phase angle which is not 90°. For simplicity $\bar{P}$ and $\bar{S}$ will be taken to have the same magnitude and to be at right angles to each other in space. As can be seen from Figure 7.8(B) the tip of the resultant $\bar{R}$ will sweep out an ellipse; similarly, it is easy to show that even when $\bar{P}$ does not equal $\bar{S}$ the resultant is elliptically polarized.

The ratio of the major to minor axes and their orientation in space depends upon the phase difference (and therefore upon the conductivity of the conductor), the relative amplitude and the spatial relationship between $\bar{P}$ and $\bar{S}$.

Deriving an Equation for the Polarization Ellipse

It is useful to derive the equation for the ellipse of polarization mathematically.

Consider the primary and secondary vectors

$$\bar{P}(t) = \bar{P}\cos\omega t; \ \bar{S}(t) = \bar{S}\cos(\omega t + \phi) \tag{7.11}$$

and let them be separated in space by angle χ, Figure 7.9(A). Let the plane containing $\bar{P}$ and $\bar{S}$ be the (x,y) plane and $\bar{R}$ be the resultant $\bar{R} = \bar{i}R_x + \bar{j}R_y$

From Figure 7.9(B)

$$R_x = P_x \cos\omega t + S_x \cos(\omega t + \phi) = X \cos(\omega t + \phi_1) \tag{7.12}$$

or

$$\frac{R_x}{X} = \cos\omega t \cos\phi_1 - \sin\omega t \sin\phi_1 \tag{7.13}$$

and

$$R_y = P_y \cos\omega t + S_y \cos(\omega t + \phi) = Y \cos(\omega t + \phi_2) \tag{7.14}$$

or

$$\frac{R_y}{Y} = \cos\omega t \cos \phi_2 - \sin\omega t \sin\phi_2 \tag{7.15}$$

Multiplying 7.13 by $\cos\phi_2$ and 7.15 by $\cos\phi_1$ and combining the two we obtain

$$\begin{aligned}(\frac{R_x}{X} \cos\phi_2 - \frac{R_y}{Y} \cos\phi_1) &= \sin\omega t \sin(\phi_2 - \phi_1) \\ &= \sin\omega t \sin\psi\end{aligned} \tag{7.16}$$

where $\psi = \phi_2 - \phi_1$

Multiplying 7.13 and 7.15 by $\sin\phi_2$ and $\sin\phi_1$, respectively and again combining, we obtain

$$\left(\frac{R_x}{X}\sin\phi_2 - \frac{R_y}{Y}\sin\phi_1\right) = \cos\omega t\,\sin(\phi_2 - \phi_1)$$
$$= \cos\omega t\,\sin\psi \qquad (7.17)$$

Eliminating t from 7.16 and 7.17 by squaring and adding,

$$\frac{R_x^2}{X^2} + \frac{R_y^2}{Y^2} - \frac{2R_xR_y}{XY}\cos\psi = \sin^2\psi \qquad (7.18)$$

7.18 is the equation of an ellipse whose major axis is inclined at an angle θ with respect to the x-axis, where θ is given by

$$\tan 2\theta = \frac{2XY\cos\psi}{X^2 - Y^2} \qquad (7.19)$$

From 7.18 it can be seen that $\bar{R}$ never vanishes but changes continuously in magnitude with the tip of $\bar{R}$ following an ellipse - the ellipse of polarization.

If $\psi = 0$, $\cos\psi = 1$, $\sin\psi = 0$, (i.e. $\phi_2 = \phi_1$) and equation 7.18 becomes

$$R_xY = R_yX \qquad (7.20)$$

which is the equation of a straight line through the origin. Therefore, for $\psi = 0$, or no phase difference, we have plane polarized waves, which conforms with Figure 7.7.

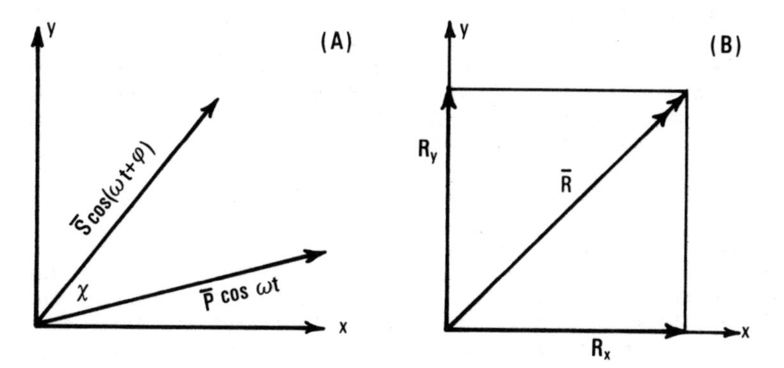

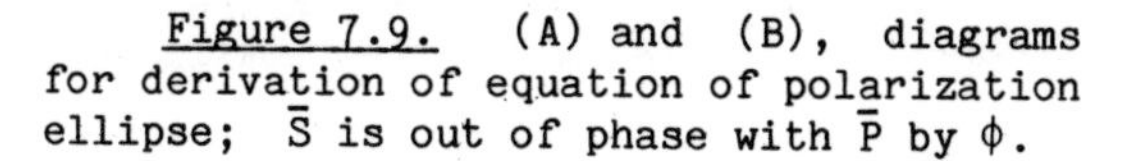

Figure 7.9. (A) and (B), diagrams for derivation of equation of polarization ellipse; $\bar{S}$ is out of phase with $\bar{P}$ by ϕ.

If $\psi = \pi/2$, $\cos\psi = 0$, $\sin\psi = 1$ and equation 7.18 becomes

$$\frac{R_x^2}{X^2} + \frac{R_y^2}{Y^2} = 1 \tag{7.21}$$

which is the equation of an ellipse with axes in the x and y direction, i.e. $\theta = 0$ or $\pi/2$. Therefore when $\bar{P}$ and $\bar{S}$ are at right angles, $\theta = \pi/2$. If in addition X=Y then

$$R_x^2 + R_y^2 = X^2 = Y^2 \tag{7.22}$$

which is the equation of a circle and we have circularly polarized waves, conforming with Figure 7.8(A).

Crude Uses of Polarization Ellipse Characteristics

The signal amplitude observed at the receiving coil, Rx, is proportional to the component of the resultant $\bar{R}$, lying along the coil axis, Figure 7.8(C). It will therefore be a maximum when the coil axis is parallel to the major axis of the polarization ellipse and a minimum when the coil axis is parallel with the minor axis, Figure 7.8(D); the ratio of the maximum signal to minimum signal will therefore give the ratio of the major to minor axes.

A secondary magnetic field which is only slightly out of phase with the primary field will produce a very narrow polarization ellipse, almost linearly polarized (see equation 7.29), and a sharp null should therefore be observed. This is similar to the earlier observation that for a good conductor the phase angle α, Figs. 7.5 and 7.6, goes to 90° and the polarization approaches linear. Therefore, the narrowness of the polarization ellipse can be used as a rough measure of the grade of a conductor.

DIPOLE APPROXIMATION

The magnetic field near a loop transmitter is very complex but becomes simpler as the distance from the transmitter increases until, at a distance of a few Tx loop diameters, the field approximates that of an oscillating dipole.

MUTUAL INDUCTANCE

The term mutual inductance is used to describe the interaction at a distance between electrical circuits due to electromagnetic induction. If two coils of small winding cross-sections are held in fixed orientation and distance from each other, the voltage induced in the receiver is proportional to the time rate of change of current in the transmitter. The constant of proportionality is called the coefficient of mutual inductance.

If $I_T e^{i\omega t}$ is the current flowing in the Tx then the voltage in the Rx is given by

$$E = -M_{12} \frac{\partial}{\partial t} (I_T e^{i\omega t}) \tag{7.23}$$

Provided the two coils remain in the same position and orientation their functions as Rx and Tx may be interchanged and the Helmholtz reciprocity law applied which states that in such conditions $M_{12} = M_{21}$. However, any variation in the electrical conductivity of the medium between the transmitter and the receiver can affect the coefficient of mutual inductance. This is indicated schematically in Figure 7.10. This schematic is meant to indicate a system with a Tx, a Rx and a subsurface conductor, C, which has inductance and resistance. M_{TR}, M_{TC} and M_{CR} are the coefficients of mutual inductance between the three electrical circuits.

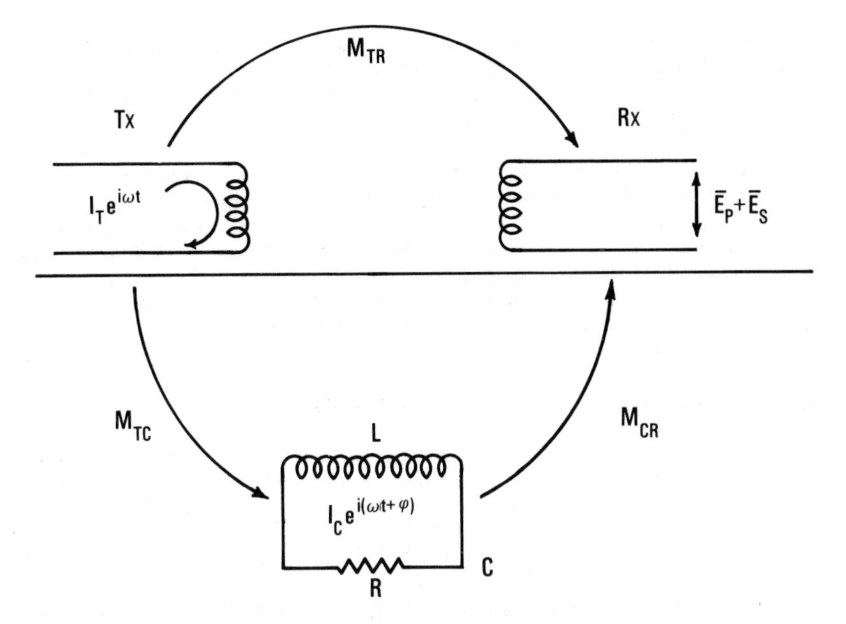

Figure 7.10. Schematic illustrating how the transmitter causes the subsurface conductor to act as a secondary transmitter, the resultant of the primary and secondary being detected at the receiver, with phase and amplitude depending on the three coefficients of mutual induction.

When switched on the Tx induces voltages, and therefore currents, in the receiver and in the subsurface conductor. The subsurface conductor then acts as a secondary transmitter which in turn induces further currents in the receiver. The total induced voltage, $\bar{E}$, in the receiver is the vector sum of that due to the primary ($\bar{E}_p$) field, which is a function of M_{TR}, and that due to the secondary field ($\bar{E}_s$) of the conductor, which is a function of both M_{TC} and M_{CR}.

In areas where there are no conducting subsurface bodies it can be assumed that M_{TC} and M_{CR} are zero and there is no secondary component; that is $\bar{E}_s = 0$. If a move is made into an area where there are subsurface conductors, that is $\bar{E}_s$ is not zero, then we get a change in the total field due to the change in M_{TC} and M_{CR}. Since these two mutual inductances, together with M_{TR}, are really the separated parts of the whole mutual inductance for the system, electromagnetic anomalies can be regarded simply as changes in the coefficient of mutual inductance.

This leads to the important observation that <u>anything</u> that alters the mutual inductance will appear as an electromagnetic anomaly, even a change in the orientation or distance between the transmitter and receiver coils.

Therefore, one of the most important things to remember about an electromagnetic method which makes use of phase relationships or intensities, particularly those methods with portable transmitters or receivers, is that the coil separation and orientation must be kept constant at all times within a given survey to ensure that the contribution due to the Tx-Rx coupling remains constant. This is not necessary for methods using the dip angle technique.

<u>Three Special Cases of Particular Interest</u>

1. When the loops, of radii r_1 and r_2, are coaxial, Figure 7.11(A), and the distance between their centres is ℓ (very much greater than r_1 and r_2) then

$$M_{12} = \frac{\mu_o \pi r_1^2 r_2^2 I_1}{2\ell^3} \qquad (7.24)$$

2. When the loops are coplanar, Figure 7.11(B), and using the same symbols as before

$$M_{12} = -\frac{\mu_o \pi r_1^2 r_2^2 I_1}{4\ell^3} \qquad (7.25)$$

This arrangement is the basis of the horizontal loop or "Slingram" method. The sign indicates whether flux linkage is such that current flow in the Rx is parallel (negative sign) or anti-parallel (positive sign) to that in the Tx.

Although the magnitude of the coupling coefficient in case 2 is only half that of case 1, both arrangements are referred to as having a maximum coupling.

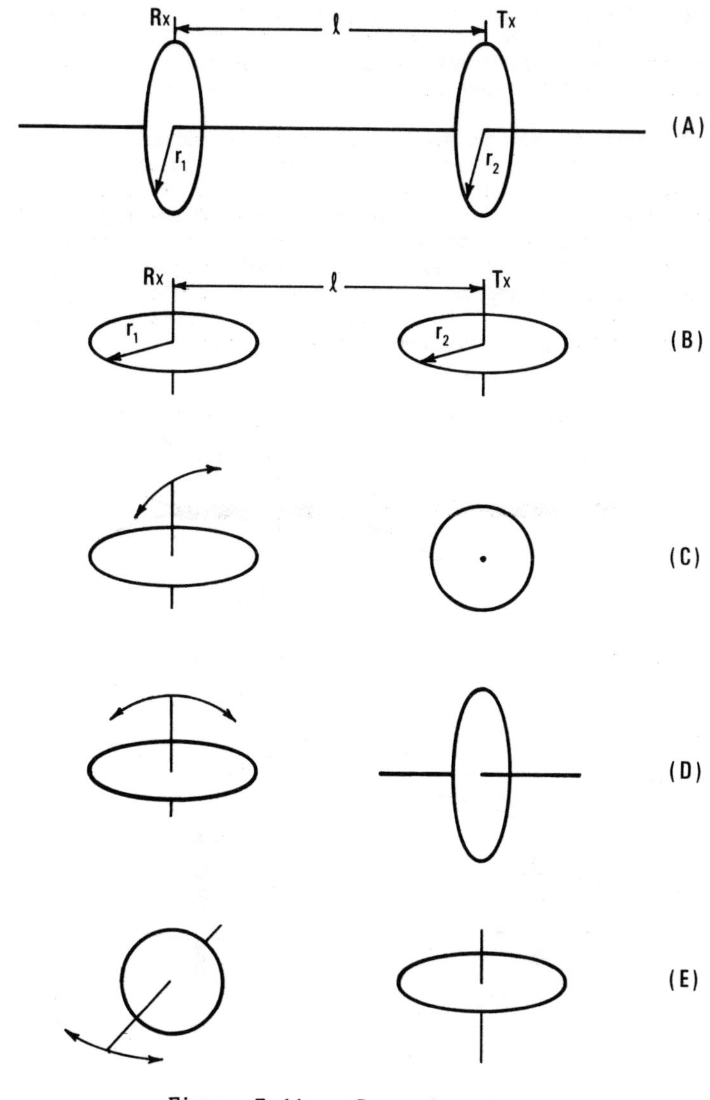

Figure 7.11. Some important Tx-Rx geometries: (A) and (B) maximum coupling positions - often used in methods using phase and intensity relationships. (C)-(E) Minimum coupling positions - often used in dip angle techniques.

3. When the axes of the two loops are orthogonal, Figure 7.11 (C)-(E), the coupling coefficent, in free space, is zero. These

arrangements are used, in the initial position, in a number of dip angle methods. The arrangement shown in 7.11(E) is the only one where rotation of the receiver in the direction indicated keeps the axes mutually perpendicular at all times.

To indicate how in some situations the mutual induction and coupling coefficient is positive whereas in others it might be negative, flux linkage diagrams are shown in Figure 7.12. The figures represent an instantaneous picture of the magnetic fields, with the transmitter (primary) and receiver (secondary) coils sufficiently far apart that a simple dipole magnetic field results and can be represented by a bar magnet, or dipole, along the axes of the coils. In both cases it is necessary to recall the definition of mutual inductance given earlier.

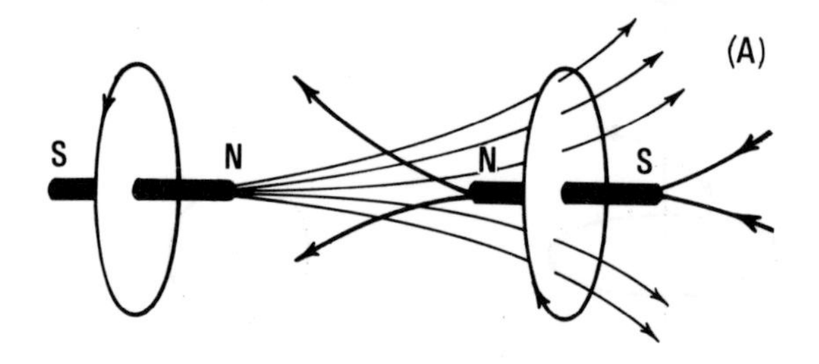

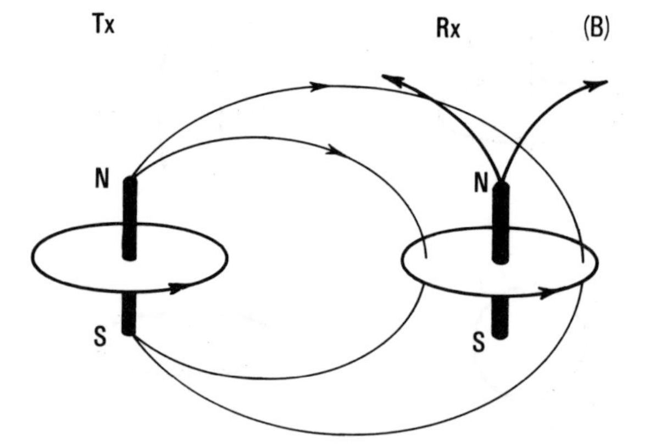

Figure 7.12. Flux linkage diagrams illustrating how change of sign of mutual inductance arises.

In Figure 7.12(A), for the secondary field to oppose the primary field the current flow in the secondary must be in the opposite sense to that in the primary. Therefore the coefficient of mutual inductance is positive as is the coupling coefficient.

In Figure 7.12(B), for the secondary field to oppose the primary field, the current flow in the secondary must be in the same sense as in the primary; in this case the coefficient of mutual induction is negative as is the coupling coefficient.

GENERAL CLASSIFICATION OF METHODS

As with most exploration methods, classification can be made in different ways. For instance, classification could be by type of motion of the Tx

(a) moving source methods
(b) fixed source methods

or by type of measurement made with the Rx

(a) angle measuring systems
(b) phase measuring systems
(c) intensity measuring systems

To avoid confusion some explanation must be given as to what is meant by "phase measuring systems".

Meaning of "In Phase", "Out of Phase", "Quadrature", "Real", "Imaginary"

Referring to the earlier phase diagram, Figure 7.6, it can be seen that the projection of the resultant, $\bar{R}$, on the OA axis is $\bar{R}\cos\phi$. Since this component is parallel to $\bar{P}$ we say that this component of $\bar{R}$ is "in phase" with $\bar{P}$. That is, $\bar{R}\cos\phi$ is the "in phase" or "real" component, Re, of $\bar{R}$.

Similarly, the projection of $\bar{R}$ on the AQ axis is $\bar{R}\sin\phi$; since it is perpendicular to $\bar{P}$, $\bar{R}\sin\phi$ is called the "out of phase", "quadrature" or "imaginary" component, Im, of the resultant.

The secondary field can also be resolved into real and imaginary components in the same way. The terms real and imaginary come from the representation of a sinusoidally varying current (or field) by $I = I_0 e^{i\omega t}$, or

$$I = I_o(\cos\omega t + i\sin\omega t) \tag{7.26}$$

$I_0\cos\omega t$ is the real part; $I_0\sin\omega t$ is the imaginary part and is clearly 90° out of phase with the real part since $\cos\omega t = \sin(90 - \omega t)$. It should also be noted that for a 180° phase difference, i.e. $\omega t=(\omega t+180)$, equation 7.26 holds but with a change of sign. When dealing with magnitudes of the Re and Im components, a 180° phase difference can be regarded as an in phase component with a negative magnitude.

Measurement of Real and Imaginary Components

To measure the relative intensities of the real and imaginary components we require a basis of reference. In fact two reference signals are required; one which is in phase and one which is 90° out of phase with the transmitted signals. These reference signals must be available at the receiver via a route that undergoes no interference due to subsurface conductors since interference from such a source would alter their phase relative to that at the transmitter.

In practice this means that the transmitter should be connected to the receiver with a cable which carries the transmitted signal, plus the 90° out of phase reference signal, directly to the receiver. Alternatively, digitised reference signals can be transmitted by radio link since although amplitudes may vary, the digits (or frequencies) do not suffer the phase effects that analog signals do.

From this point there are a large number of alternative measurement possibilities, nearly all of which have been used in some form or another, which lead to competing claims as to which is the "best" method.

Measurements can be made of the real and imaginary components of

(a) the horizontal component of the resultant
(b) the horizontal component of the secondary
(c) the vertical component of the resultant
(d) the vertical component of the secondary
(e) the total field of the resultant
(f) the total field of the secondary

Secondary Field Relationships

Since the ratio of the strength of the real components (Re) to that of the imaginary component (Im) is $S \sin\alpha / S \cos\alpha = \tan\alpha$, and since α is a function of the conductivity, σ, the ratio of Re to Im should give some information about the conductivity. This might be seen from Figure 7.13 and Table 7.1.

In Figure 7.13(C) a definite statement about the secondary cannot be made without specifying a phase angle but the important point to note is that in 7.13(A) and (B) the amplitude of the imaginary component tends to zero but in Figure 7.13(C) it can be seen that it must have a finite magnitude; therefore it can be stated that in going from a good conductor ($\alpha \to 90^{o}$) to a poor conductor ($\alpha \to 0^{o}$), Im will pass through a maximum.

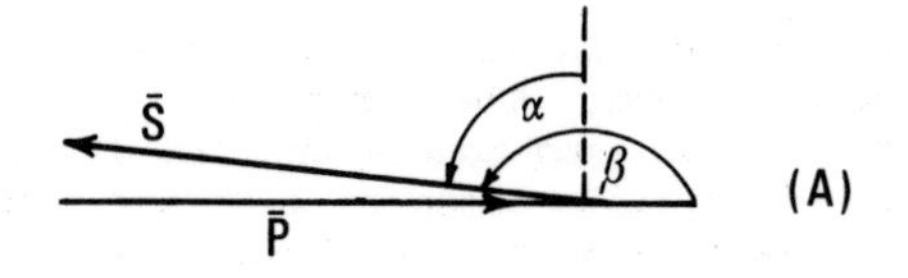

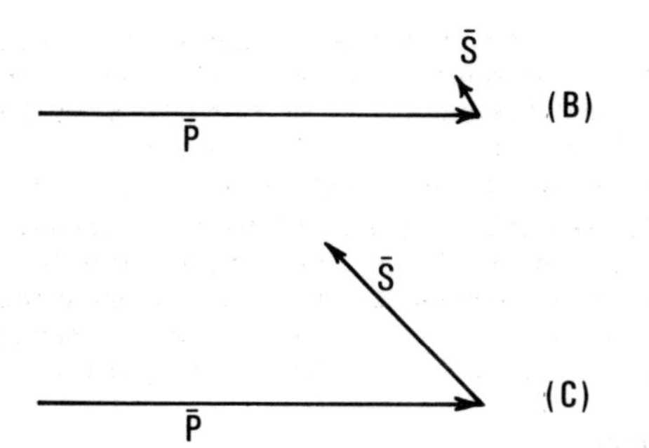

Figure 7.13. Secondary field relationships for (A) good, (B) poor, (C) intermediate conductors.

Table 7.1

Summary of information content on anomalous conductivity(σ) in secondary magnetic field

σ	α	Re	Im	β
$\to\infty$	90	large	0	180
$\to 0$	0	0	0	90
intermediate	intermediate	intermediate	Passes through a maximum	intermediate

Propagation Constant

The propagation of electromagnetic waves through a semi-infinite medium depends on the propagation constant

$$\gamma = \sqrt{\mu\varepsilon\omega^2 + i\mu\sigma\omega} \quad (7.27)$$

The first term, the real part, refers to the displacement current while the second term, the imaginary part, refers to the conduction current. In good conductors the displacement currents are completely negligible compared with the conduction currents. In typical earth materials it is found that even in normally nonconducting rocks the displacement currents are at least three orders of magnitude less than the conduction current at the frequencies used with typical field equipment (around 1 kHz).

It will be noted that the propagation constant depends upon the magnetic permeability as well as upon the electrical conductivity (and frequency). By far the most important physical property, because of its wide variability, is the electrical conductivity although there may be some very special cases where the magnetic permeability should be considered. Another point to bear in mind is that it has been implicitly assumed that conductivity and permeability are independent of the frequency being used; this is not necessarily true although once again, the cases in which this is likely to become important are quite rare.

Induction Number

In a real situation the response will depend on geological structure and thus on the geometry of the subsurface bodies and Tx-Rx separation. We find it convenient to express the response in terms of a dimensionless parameter ϕ, where

$$\phi = (\sigma\mu\omega)^{\frac{1}{2}}\ell \tag{7.28}$$

and is called the induction number; the terms in brackets have their usual meanings but ℓ is a geometrical term which has dimensions of length. More often, the term $\theta(=\phi^2)$ is used in which case it is called the response or induction parameter; this is equivalent to the response parameter $\alpha = \tan^{-1}(\omega L/R)$ in the purely electrical case.

Response parameters for typical bodies in a uniform field are shown in table 7.2.

TYPICAL EXPLORATION PROBLEM

A typical geological situation is indicated in Figure 7.14. We wish to determine and interpret the nature of the total magnetic field $\bar{H}_T$ at the receiver which results from the vector addition of the source field, $\bar{H}_p$, the secondary field $\bar{H}_o$ of the air filled half space above ground, the secondary field of the overburden, $\bar{H}_1$, the secondary field of the target $\bar{H}_3$, and the secondary field of the material making up the host rock which, for simplicity, is combined and called $\bar{H}_2$. Ideally, from this total vector summed field of $\bar{H}_T = \bar{H}_p + \Sigma H_i$ we would like to find out something about σ_i, ε_i and μ_i (i=0,1,2,3).

Table 7.2

Typical response parameter in a uniform field

Geometry	θ	Symbol meaning
Sphere	$\mu\sigma\omega a^2$	a = radius
infinite horizontal cylinder	$\mu\sigma\omega a^2$	a = radius
thin disk	$\mu\sigma\omega ta$	a = radius t = thickness
thin tabular body (high dip angle)	$\mu\sigma\omega Ws$	W = thickness s = Tx-Rx separation

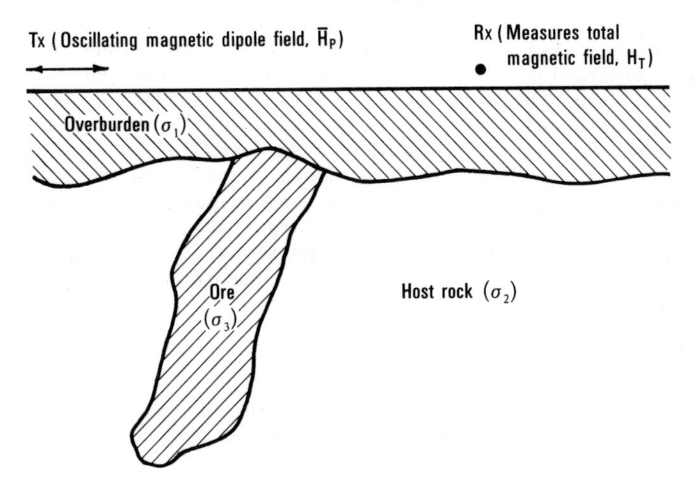

Figure 7.14. A typical exploration problem with steeply dipping conductor buried beneath overburden.

This is obviously a difficult problem so it is simplified by taking $\sigma_o = 0$ and assuming that σ_1 is small compared with σ_3; that is $\overline{H}_o$ and $\overline{H}_2$ are ignored.

In some cases, but by no means all, $\overline{H}_1$ can also be ignored.

This is essentially a practical problem requiring the matching of the system used to the conditions underground. Clearly, if there is a conductive overburden the transmitter-receiver geometry used should minimize the coupling with horizontal sheets. If the target conductors are also horizontal the exploration problem becomes intractable. Fortunately, in the Canadian shield the ore usually forms as elliptical lenses or steeply dipping dykes; of course, it could also be that horizontally oriented deposits have not been located.

TYPICAL TRANSMITTER-RECEIVER GEOMETRIES

In Figure 7.11, when the Tx and Rx are arranged as in (A) and (B), they are known as symmetrical systems; this means that the coils are maintained in a fixed relative position so that if they are constructed to be both transmitters and receivers, the direction of traverse makes no difference in the readings; this follows from the reciprocity for mutual inductance whereby the source and receiver can be interchanged. The data are therefore much easier to interpret.

In particular, arrangement (B) is much used in a system known generally as "Slingram". It is most important that the coils be maintained parallel to each other, in the same plane and with constant separation because it should be remembered that it is the mutual inductance between the two coils and subsurface body that is being measured and this can be significantly affected by coil misalignment or spatial variations. For example in the typical Slingram configuration, with a normal separation of 100 meters, the in-phase component changes by approximately 5% per meter change in separation. This rather large variation occurs because the simplifying use of the dipole approximation also means that the magnetic field decreases as the cube of the coil separation (see equations 6.10 and 6.11).

TYPICAL FIELD ARRANGEMENTS

To appreciate how the general form of an anomaly curve arises, flux linkage diagrams are drawn for some typical and much used geometries.

Tilt Angle Techniques

These are essentially reconnaisance techniques used to establish quickly the strike of a conductor by observing where the sign of the angle of tilt of the resultant polarization ellipse changes sign; however, some crude quantitative estimates of conductor quality and geometry can also be obtained.

In-line technique. There are a number of different geometries but that illustrated in Figure 7.15 is for a vertical Tx with its axis pointing in the direction of motion and of the

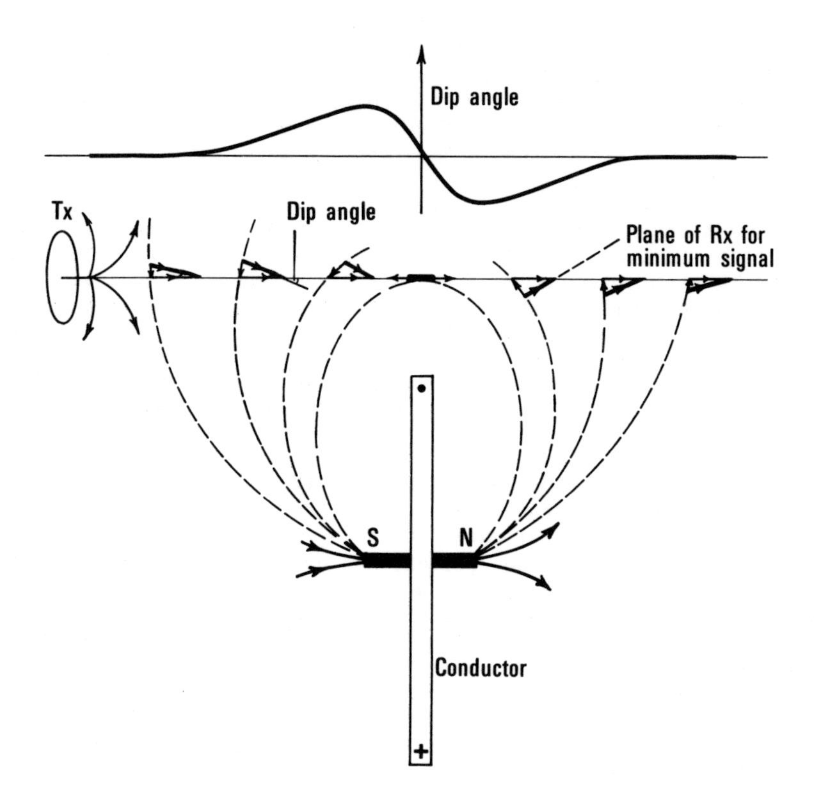

Figure 7.15. Illustrating origin of "cross over" for inline dip angle method using vertical Tx. If Tx-Rx separation is constant, the primary field, which is horizontal, is always of same amplitude at Rx. Spot on conductor indicates eddy current flowing out of page, while + indicates flowing into page. Dip angle is measured by selecting a reference level and direction and is negative when smallest angle (<90^{0}) has to be measured anticlockwise and positive when measured clockwise.

Rx. Since the Tx is vertical, its axis and therefore its equivalent magnetic dipole is horizontal.

In the field situation the Tx-Rx distance is maintained constant so that for the moving Tx-Rx system the primary field has a constant amplitude at the receiver; the only variable will therefore be the secondary field. This is illustrated with an instantaneous picture in Figure 7.15. It is therefore the variation in the secondary field, and consequently the resultant, which gives information about variations in subsurface electrical conducting properties. From Figure 7.15 it can be seen that the amplitude of the secondary field decreases in amplitude and increases in tilt as the Tx-Rx system is moved away from the ore body.

The tilt or dip angle must be measured according to some direction (usually the forward direction) and reference surface; if the levelling device is a bubble the reference surface is the gravitational equipotential surface which, for most practical purposes is parallel to the earth`s surface. In Figure 7.15 if the direction of the primary arrow is taken as positive, then the tilt angle is negative when the smallest angle (less than 90°) has to be measured anti-clockwise and positive when the smallest angle has to be measured clockwise.

It is difficult to show quantitatively how the maximum and minimum arise in Figure 7.15 because the resultant depends upon both the magnitude and orientation of the secondary, and near the orebody they change too rapidly to be shown clearly in Figure 7.15. Nevertheless, it is obvious that maxima and minima must occur since it is readily seen that at positions well away from the orebody the resultant field has zero dip and increases as the orebody is approached, yet when the receiver is directly over the body in a position of minimum coupling, the resultant field again has zero dip; therefore, somewhere between these two regions the dip angle amplitude must peak. Similarly, on the other side of the body there must be a minimum.

Broadside technique. In another technique the transmitter and receiver are again kept at a constant distance but are advanced in a broadside and not an in-line manner, Figure 7.16. It is difficult to show the flux diagrams in three dimensions, so the azimuthal component is ignored, as it often is in practice, and only the component of interest is shown. It should be noted that in this case the sense of the field lines are different from those in Figure 7.15 because the transmitter axis is not pointing towards the receiver although it is pointing in the direction of traverse. In this case the resultant field is such that the plane of the receiver at minimum dips away from the target.

The coil arrangement and typical anomaly results are indicated in Figure 7.17. The dip angle is plotted against the position of the receiver but in doing this it must be remembered whether the transmitter is on, for instance, line 1 or line 3 when plotting the tilt angle for line 2.

The E-W dip of the magnetic field can be found by a second rotation and the two measurements combined to give the space orientation of the polarization ellipse; occasionally the E-W dip is found first but only the N-S dip is recorded since this is the

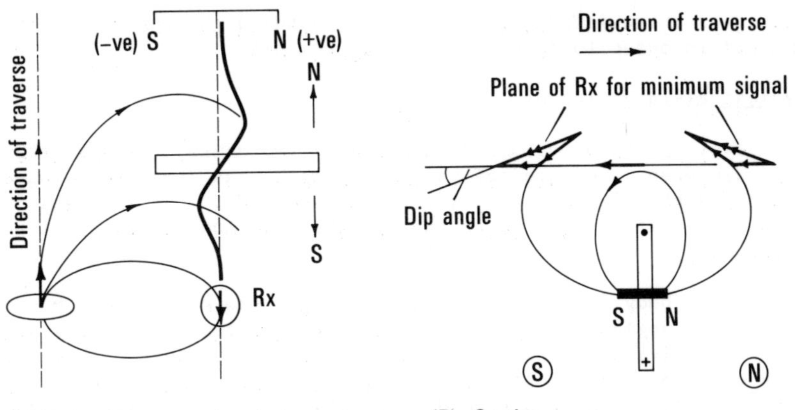

(B) Section

Figure 7.16. Illustrating origin of "cross over" for broadside dip angle method using vertical Tx. If Rx-Tx separation is constant, the primary field, which is horizontal, will have the same amplitude at Rx.

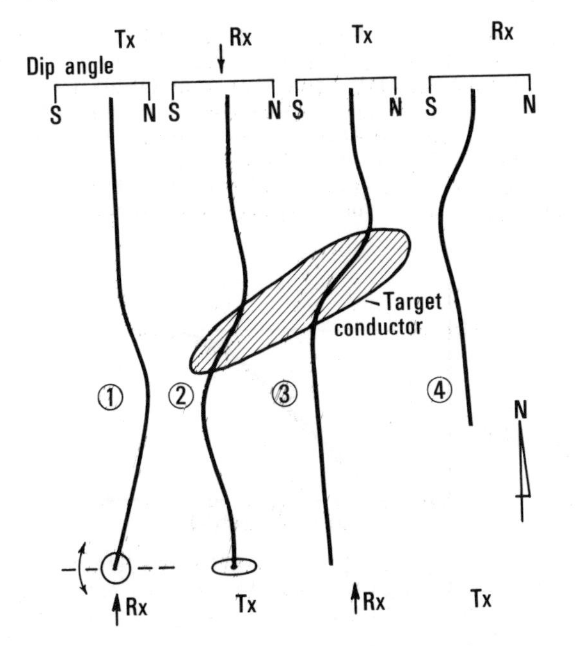

Figure 7.17. Typical field results using broadside arrangement illustrated in Figure 7.16. Receiver and transmitter functions are interchangeable so that curves for a pair of lines, say 1 and 2, can be obtained from one traverse. Anomaly curve for line 2 is obtained when Tx is on line 1.

most definitive information. However, finding the strike is often omitted in order to speed up the measurements.

Typical Field Equipment

Typical field equipment consists of two coils, approximately 1m diameter, capable of being operated at two frequencies separated by a factor of 4, say 400 and 1600 Hz. The coil spacing should be at least twice the expected maximum thickness of the overburden.

For a body of specific size and shape, the ratio, say M, of the dip angles at the low and high frequencies, will vary with the conductivity of the subsurface body in the manner shown in Figure 7.18. From the figure it can be seen that for low conductivity the higher frequency will give a much larger response than the lower frequency, whereas for high conductivity the ratio approaches one. This is because, other things being equal, frequency affects the phase angle which in turn influences the orientation of the ellipse in space (see section on theoretical development of equation of ellipse of polarization).

That this is so might be seen be referring to equation 7.9 where the phase angle $\alpha = \omega L/R$. For good conductors $R \to 0$ and $\alpha \to \pi/2$ and therefore remains essentially constant no matter what the frequency is. For poor conductors $R \to \infty$, and $\alpha \to 0$; therefore for a large body of intermediate conductivity α increases as ω increases since $\omega L/R$ is a finite number. Thus, with a two frequency system, the denominator of the abscissa in Figure 7.18 remains constant while the numerator increases with increasing frequency.

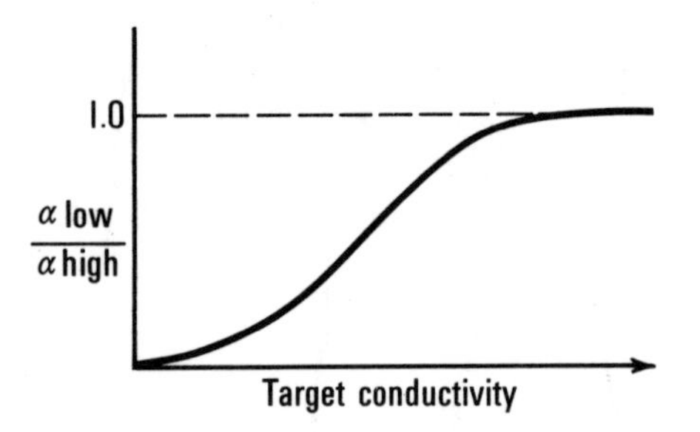

Figure 7.18. Illustrating how ratio of tilt angles may be used to estimate whether anomalous body is a good or bad conductor.

Phase Intensity Measuring Systems.

Horizontal Loop or Slingram Method. Typical results for the horizontal loop method are shown in Figure 7.19. Simple estimates of conductor width may be made in terms of the transmitter-receiver separation; the case shown in Figure 7.19 is that of a vertical, semi-infinite depth extent, infinitely long dyke with one surface coincident with the surface of the earth and with the Tx-Rx traverse perpendicular to the strike of the conductor. In practical terms it is found that if the system is three or more Tx-Rx distances away from the ends of the dyke, it may be regarded as an infinitely long dyke; furthermore, one of the advantages of the Slingram method is that although the anomaly curves shown in Figure 7.19 are for a system moving at 90° to the strike of the conductor the shape of the response curve would not be significantly different if the direction of traverse was at, say, 60° to the strike.

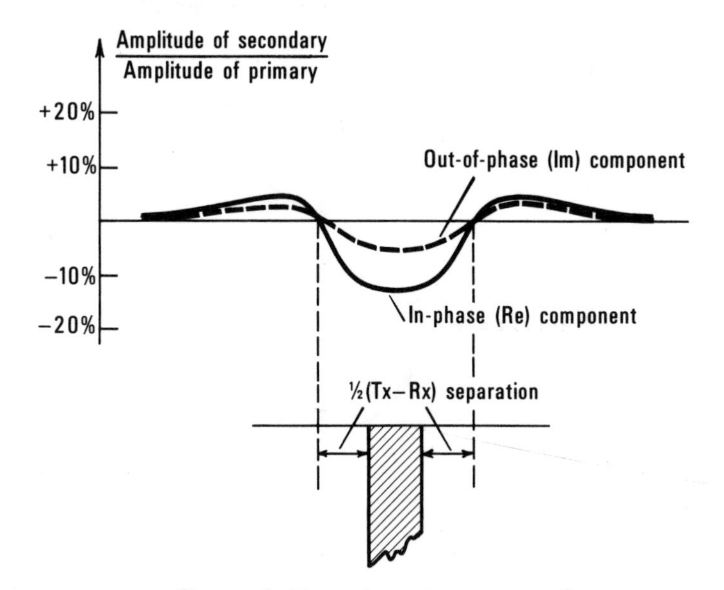

Figure 7.19. Anomaly curves for a semi-infinite conductor with its top under only thin cover using the Slingram (horizontal coplanar loops) method.

The ratio of the in-phase to out-of-phase component is large over a good conductor and small over a bad conductor (see section on secondary field relationships).

The changes in sign as the system moves over the conductor may be explained with the aid of the flux linkage diagram shown in Figure 7.20. In Figure 7.20(A) the coupling coefficient is arbitrarily given a positive sign. As the receiver moves over the orebody, since the plane of the receiver is now perpendicular to the plane of the conductor, Figure 7.20(B), there is minimum coupling between the receiver and the conductor; therefore the secondary field due to the currents induced in the conductor by the transmitter does not induce any currents in the receiver, and the ratio of the secondary field to the primary field is zero.

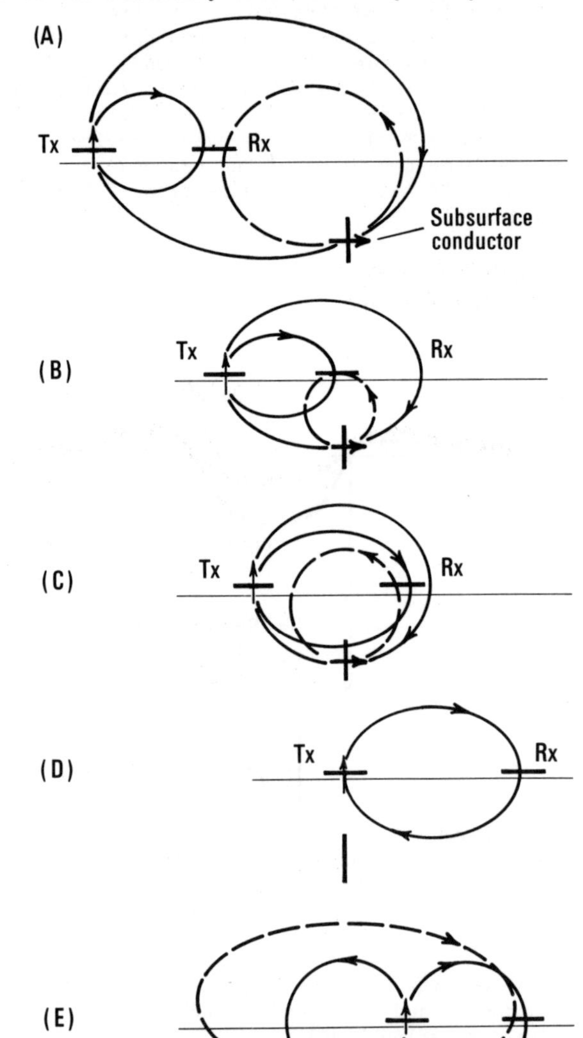

Figure 7.20. Flux linkage diagram to illustrate origin of the shape of the anomaly curve shown in Figure 7.19. Full line - primary field; dotted line - secondary field.

When the transmitter and receiver are on opposite sides of the subsurface conductor, Figure 7.20(C), it can be seen that the coupling coefficient has changed sign (see Figure 7.12) from what it was before, so that the sense of the secondary field changes sign with respect to what is was in 7.19(A).

When the geometry is such that the transmitter is directly over the conductor, the coupling coefficient between it and the conductor is zero so that no currents are induced in the underground conductor. Therefore no secondary field can be produced by the underground conductor at the receiver, and the ratio of the secondary field to the primary field is again zero.

When the system is completely on the other side of the subsurface conductor, compared with its initial position, the coupling coefficient again changes sign, as can be seen from Figure 7.20(E), so that the sense of the secondary field again becomes positive.

Clearly this is an over simplification of the real case since in general the subsurface conductors are of finite width, are not of infinite depth nor vertically oriented.

If the body has a dip between 30° and 90° the in-phase and out-of-phase anomaly curves become asymmetrical, the asymmetry being such that the areas contained between the curves and the horizontal axis are greater in the region over the body, as illustrated in Figure 7.21. Simple rules have been developed for estimating the degrees of dip from the degree of asymmetry.

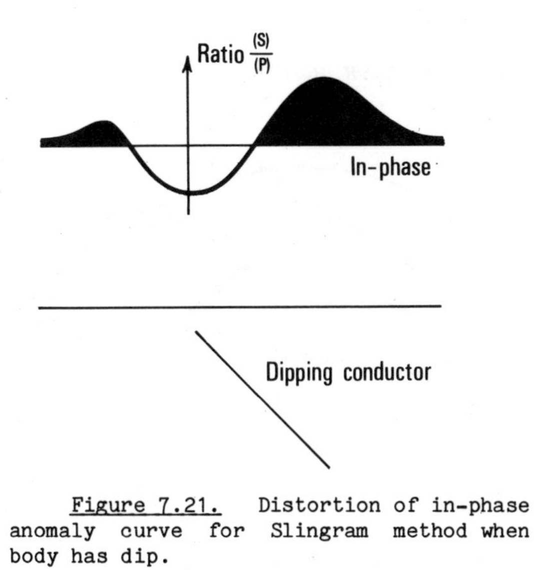

Figure 7.21. Distortion of in-phase anomaly curve for Slingram method when body has dip.

As with most electromagnetic methods, a proper interpretation should really contain a comparison of the complete field anomaly curve with either a laboratory model experiment or theoretically computed curves. However, certain short cuts can be taken using particular properties of the anomaly curve. For instance, even when the conductor dips there is still a large degree of symmetry in the anomaly curves and two easily identified properties are the maximum amplitudes of the in-phase and quadrature anomalies; these occur mid-way between the two zeros if the conductor is vertically oriented. A series of Phasor diagrams can be constructed using dimensionless ratios which involve the ratio of conductor width (W) to vertical depth (z) to the top edge of the conductor and the response parameter $\mu\sigma\omega Ws$, where s is the Tx-Rx separation and the other symbols have their usual meanings. Sets of curves for dips of 90° and 60° are shown in Figure 7.22; similar diagrams can be drawn for different dips but, as can be seen from Figure 7.22, the curves are relatively insensitive to dip and it is not really necessary to have increments of dip less than 30°.

To use the Argand diagrams, the point represented by the maximum in-phase and quadrature responses is plotted on the Argand diagram which most closely fits the dip estimated by the method illustrated in Figure 7.21. The conductivity-thickness parameter, σW, and depth to the top edge of the conductor, can immediately be found. Without an independent estimate of one of the parameters it is not possible to go further. Fortunately, it is usually possible to take a guess at the conductivity or the thickness of the subsurface conductor.

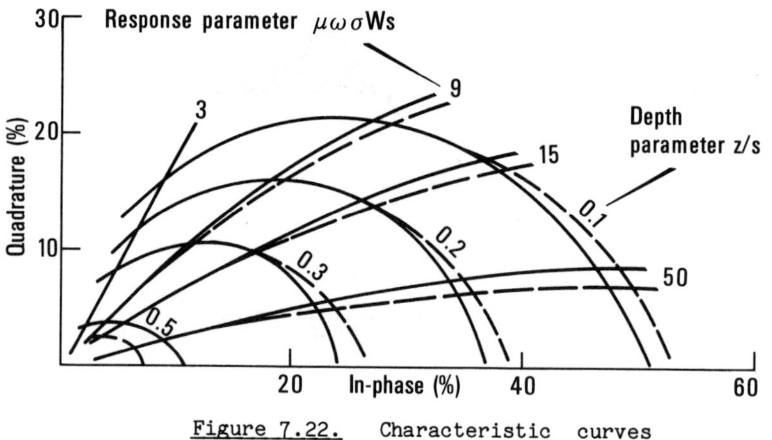

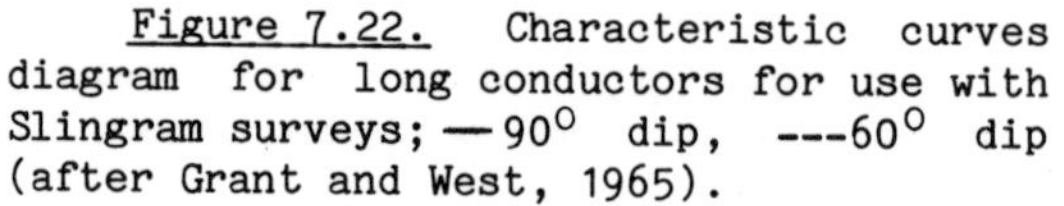
Figure 7.22. Characteristic curves diagram for long conductors for use with Slingram surveys; — 90° dip, --- 60° dip (after Grant and West, 1965).

It should be noted that in the case of a shallow, vertically oriented conductor, the width may be estimated as shown in Figure 7.19 so that the conductivity can then be obtained directly.

Care should be taken to use a self consistent set of units in substituting the various quantities into the response parameter; many incorrect estimates are made because of a careless use of units. For instance ω is circular frequency, not frequency in Hz.

The "Turam" (fixed source) method. A typical layout for the Turam method is shown in Figure 7.23.

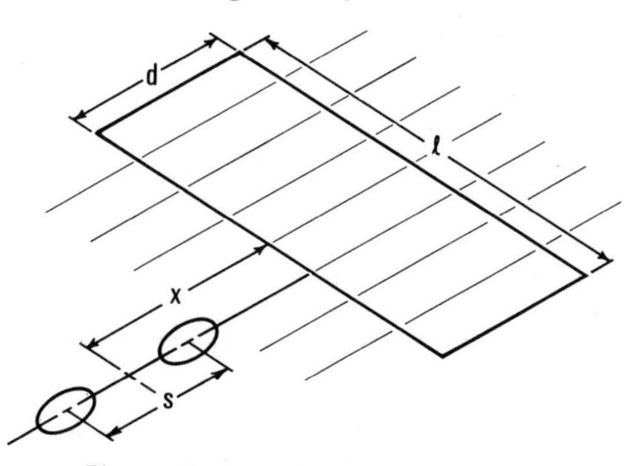

Figure 7.23. Schematic for Turam, fixed transmitter, method. Dimensions are such that d may be up to a few hundred meters; ℓ from 1 to 10 times d; x up to d; s about 0.02d to 0.1d. Measurements are rarely made within the Tx loop. Choice of dimensions usually depends mostly on depth of current penetration required but ℓ is typically 0.5 to 1 Km.

In this method the interest lies in either the field intensity variations or phase angle variations between pairs of positions on a grid. The ratios are measured by means of two small secondary coils with a separation that is small compared with the dimensions of the Tx loop; the measurements are made at grid positions separated by distances which approximate the distance between the secondary coils starting close to one edge of the primary loop and working outwards to a distance which is approximately the same as the width of the primary loop; measurements are rarely made inside the loop.

In general it is only the ratio of the vertical components that is measured, i.e. the secondary coils are kept horizontal, although occasionally horizontal components or a combination of vertical horizontal components are used.

Since the intensity falls off with distance a typical curve would appear as shown in Figure 7.24. As is the case with so many geophysical methods the appearance of the curve is much easier to understand when the residual anomaly is plotted so that it is usual to remove the normal fall-off of intensity with distance from curve A to produce curve B.

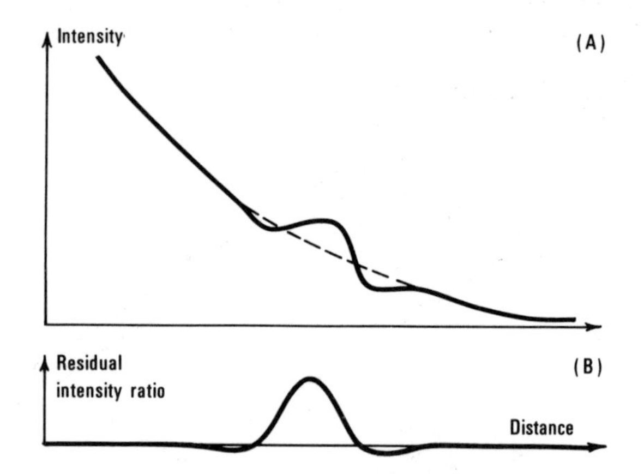

Figure 7.24. Fall of intensity with distance and residual intensity ratios (after Grant and West, 1965).

Interpretation is usually carried out on the intensity and phase difference variation. Two big advantages of this method over a moving source method are that (a) topography does not have an important effect on the results and (b) there is a constant coupling between the transmitter loop and the subsurface conductors so that the curves plotted allow accurate determinations to be made of the location at depth of the principal current axis in the subsurface. The conductivity-thickness, σW, values can be determined from the amplitude relations with the aid of Argand diagrams computed in a similar manner to those for the moving source method.

DEPTH OF PENETRATION AND MATCHING SYSTEM TO GEOLOGICAL STRUCTURE

There are many competing claims in the literature concerning depth of penetration for different techniques and even for different sets of equipment using the same technique. This arises because for a simple geometry (for example, a semi-infinite, homogeneous, isotropic medium) we can follow the physicists and define the depth of penetration (or "skin" depth) as the depth at which the field intensity is 1/e of the field intensity at the surface; for a given medium the skin depth varies inversely with $\sqrt{\omega}$ so that the higher the frequency, the smaller the skin depth.

However, the geophysical problem is much more complex and it is therefore useful to have a working definition of depth of penetration; for this purpose it is assumed that the useful depth of penetration is the maximum depth at which the response of conductors of economic interest can be clearly distinguished from the responses from other sources.

In this context then, one of the most important factors in planning an electromagnetic survey is to try to match the system chosen to the geological problem to be encountered. Sometimes, an operator has no idea what the geological problem might be in which case his first choice should be for a unit capable of rapid reconnaisance. Once the particular problem is better identified, then a choice of the technique and actual type of equipment can be made. The process of selecting a system with appropriate characteristics to match the geological problems is sometimes known as "focusing". Much of the discussion follows the work of Bosschart and Seigel.

Figure 7.25 illustrates the problem for a typical field situation and using a moving source, horizontal loop system. It is not possible to change μ, σ or W in the response parameter, $\theta = \mu\sigma\omega Ws$, since these are properties of the geology; the only way we can vary θ is to adjust ω and s, the coil frequency and separation respectively.

Clearly, if the frequency and transmitter-receiver separation is chosen such that the response parameter is towards the right hand edge of the figure, the responses of conductors with a wide range of conductivity-thickness ratios will be very similar and it will therefore be difficult to differentiate mineralized regions from host rock or overburden. On the other hand, if the operational frequency and transmitter-receiver separation are chosen so that the response parameter is in the 1 - 10 range of the figure, there will be considerable differences in response between the mineralised areas and the others. It is important to note that although changing the Tx-Rx separation changes the response parameter, making the separation large does not necessarily imply great depth of penetration in the sense that the arrangement is then more sensitive to deep conductors; it may be that the large separation simply moves the response parameter into a range where the system is sensitive to changes in the host rock or overburden conductivity-thickness parameter but insensitive to the target σW parameter.

Nearly every commercial equipment has at least two frequencies available, usually separated by a factor of about four. Few phase measuring systems have facilities for varying the separation of the transmitter and receiver since to do so requires recalibration of the reference signal so that the only readily available variable in the field is the frequency. However, recently there have appeared some sets of equipment that are

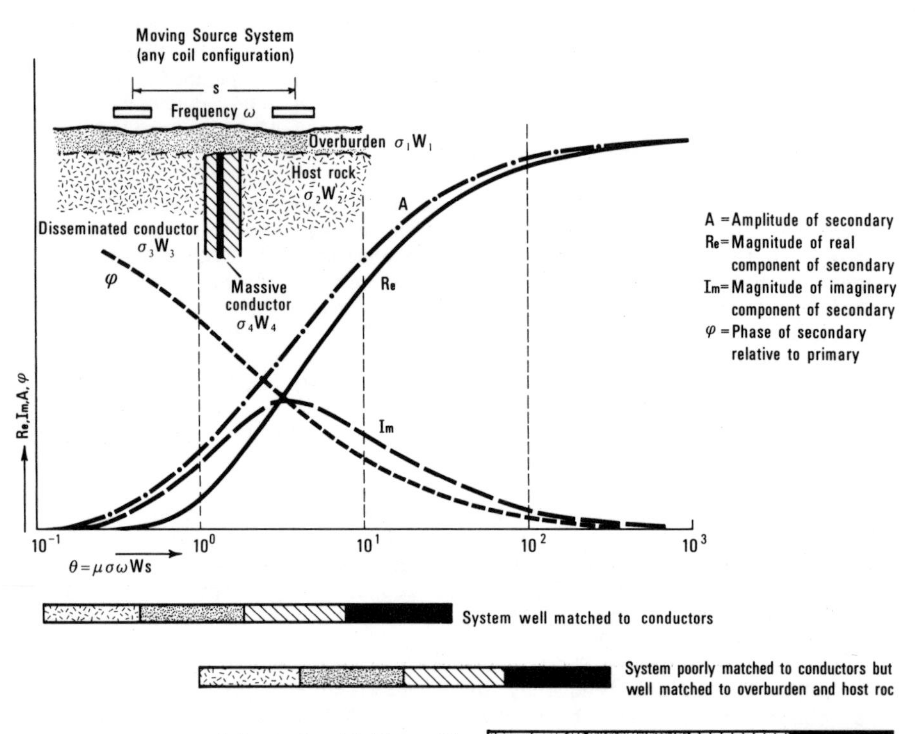

Figure 7.25. Illustration of the principles of proper matching of a system to the geological problem. The response parameter for the target (mineralisation) must fall in the sensitive portion of the Re and Im curves (after Bosschart, 1968).

capable not only of being used at a number of different separations with individual reference signals for each separation, but also possess up to half-a-dozen frequencies.

Comparison of Responses between Moving and Fixed Source Systems

Figure 7.26 indicates the type of responses which might be expected from orebodies of a given strike length using a moving source and a fixed source system. From the figure it can be seen that for conductors which have a strike length up to about 100 meters, moving source methods have the same type of response as the fixed source method. After this the field source responses increase significantly as the strike length increases while the moving source response reaches a maximum at about 200 meters and

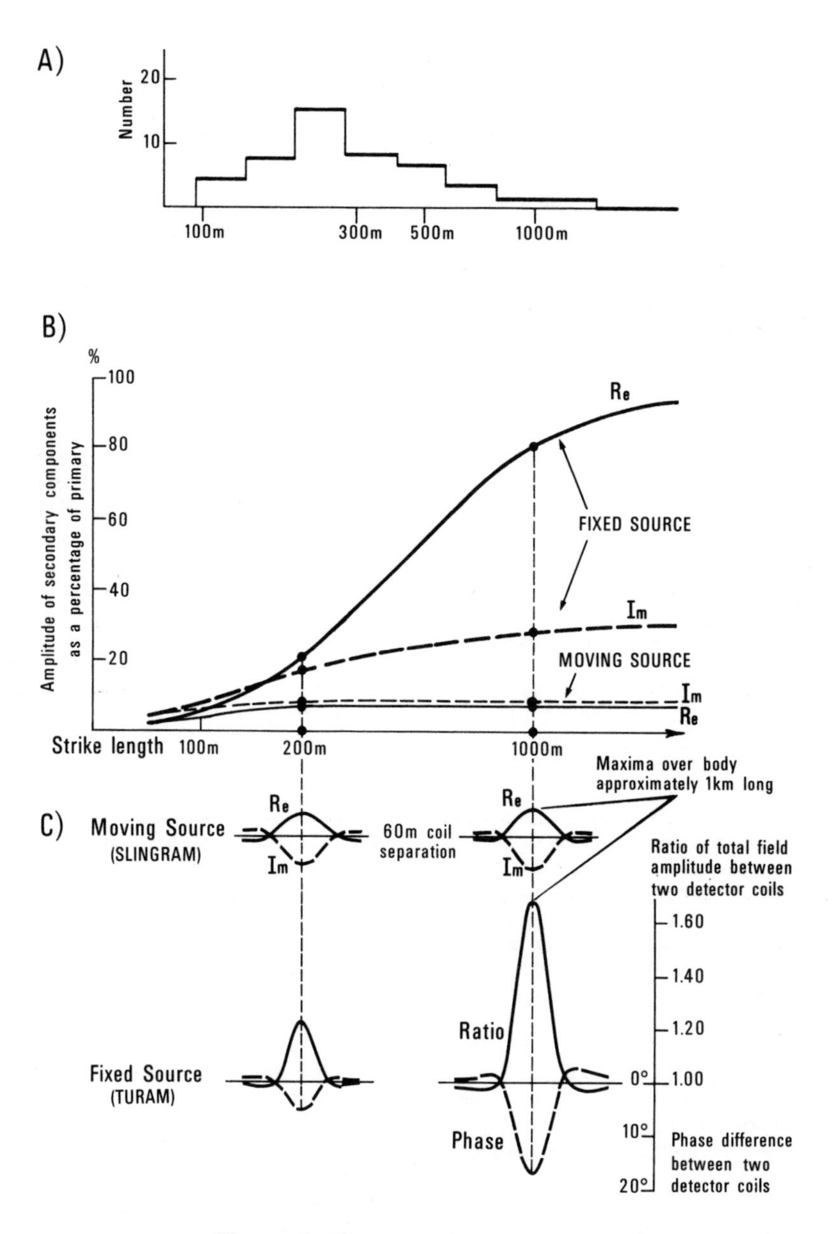

Figure 7.26. Illustrating relative effectiveness of moving source and fixed source methods with bodies of different strike length (after Bosschart and Seigel, 1966).

remains practically constant thereafter. In other words, for a moving source system the size of a conductor has relatively little effect on the magnitude of the response. Since base metals are generally small bodies the horizontal loop electromagnetic method is frequently used for them.

With the fixed source methods, although varying the size of the transmitting loop is considerably more cumbersome than varying the transmitter-receiver separation in a moving source system, it is effective in increasing the signal to noise ratio in the presence of a conductive overburden. The effect of a conducting overburden can also be minimized by a proper choice of frequencies.

VERY LOW FREQUENCY (VLF) SYSTEMS

These are essentially orientation measuring receivers using the carrier waves of pre-existing and powerful transmitter stations operating at frequencies around 20 kHz. Compared with the frequencies discussed earlier 20 kHz is high rather than low; however, the transmitting stations are used for armed services communications and in terms of communications frequencies this is very low. For instance compare this with the MHz or hundreds of MHz frequencies in the commercial AM and FM bands.

The area to be surveyed is usually far, of the order of hundreds to thousands of kilometers, from the transmitter stations so that the primary field is uniform in the nonconducting area. This makes the theoretical prediction of anomalies relatively simple.

For a survey it is usual to choose a transmitting station which yields a magnetic field which is perpendicular to the suspected strike of the target, e.g. a conducting dike, that is the strike should be close to the direction of the Tx from the field site and the primary field parallel to the traverse direction, Figure 7.27. This arrangement provides maximum coupling of the transmitted field with the orebody.

The conducting body produces secondary fields which, because the primary field is uniform, tend to flow in such a way that it forms line sources concentrated near the outer edges of the dike. The dike is therefore effectively replaced by a loop of wire whose dimensions approximate those of the dike.

Anomalies are defined by tilt angle cross overs where readings of the tilt angle of the polarization ellipse go from positive to negative. The results can be plotted directly in degrees or, since the tangent of the tilt angle is approximately the ratio of the real component of the secondary field to the primary field, as tangents of the tilt angle expressed as a percentage in a manner similar to that for the Slingram systems;

the quadrature component can also be found. It is as well to point out that even orientation measuring systems should be matched to the problem.

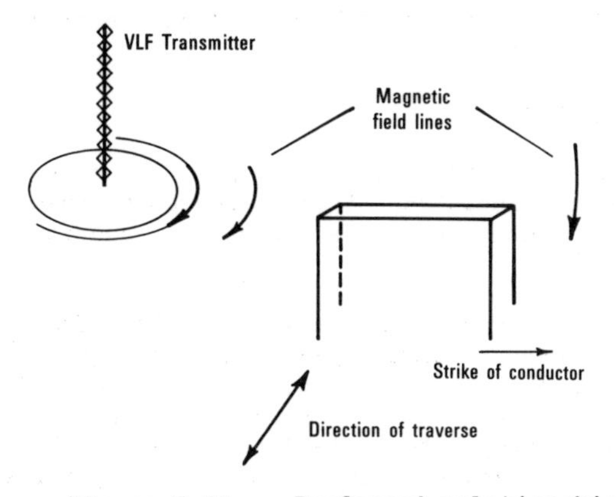

Figure 7.27. Preferred relationship between VLF transmitter, strike of conductor and field traverse.

It might be wondered why regular commercial station carrier bands are not used; this may be understood by referring to the brief discussion on skin depth.

AIRBORNE MEASURING SYSTEMS

Many of the ground based techniques can be adapted for use in the air, but because of the much more rapid coverage of ground and the greater carrying capacity of aircraft than the human beast, it becomes worthwhile devising much more sophisticated systems. For this reason there are many variations on the same basic theme - namely, the detection of anomalous secondary induced components. Therefore, in these last sections we will confine ourselves to a brief description of general problems and a few of the more unusual approaches. Many of the names used are patented and reference to a particular manufacturer's equipment has no implication that this system is better than a competitor's similar one.

General Classes of Methods

The airborne methods can be broadly classified into two main groups (I) Rigid Boom, and (II) Towed Bird, Figure 7.28.

In group I the separation and orientation of the Tx and Rx coils are rigidly maintained constant. This may be done by mounting coils directly on the aircraft, on the wing tips or fore and aft, or by towing a "bird" in which both the Tx and Rx are mounted.

In group II, the Rx is towed up to a few hundred meters aft of the aircraft with the Tx coil mounted on the aircraft or on the ground.

Some General Problems

Most of the problems which have to be solved arise from changes in mutual inductance from causes other than variations in subsurface conductivity (see earlier section on Mutual Inductance).

Some of the sources of noise are (a) stray transient magnetic fields from aircraft engine parts such as magnetos, spark plugs etc. These sources only become important if the Rx is closer than about 100 m; if the Rx is mounted on the aircraft, special precautions for appropriate shielding have to be taken. (b) Rotation of the system about the towing axis, tilts relative to the ground for the rigid boom class plus separation changes for the towed bird class; in general, rotation should be kept to 2^0 and separation maintained to 1 part in 5000 so that good design of "birds" is necessary. (c) Motion of metallic parts of the plane. This is particularly troublesome if the coils are asymmetrically located relative to the axis of the plane, or if aircraft metal parts move asymmetrically (e.g. wing or rotor motion); these problems can be avoided by using non-metallic planes.

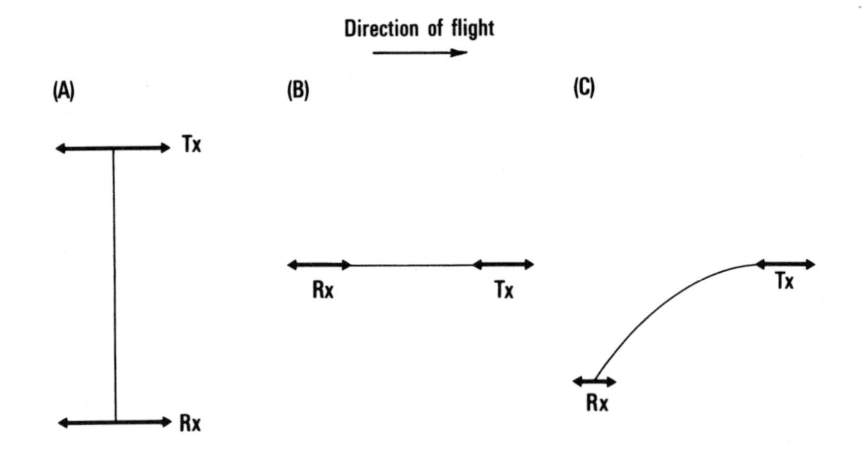

Figure 7.28. Typical geometries for airborne electromagnetic surveys (A) Rigid Boom, Broadside, (B) Rigid Boom, In-line; (C) Towed Bird, In-line.

Some Airborne VLF Systems

All these systems give only shallow penetration and are used principally for detecting gravel or sand deposits, industrial minerals etc.

Barringer Radio-phase. Experiments have shown that the vertical VLF electric field (see Figure 7.27) is stable to within about 1% with respect to the field propagated in free-space regardless of changes in topography and conductivity of the underlying material. This system therefore measures the in-phase and out-phase components of the horizontal magnetic field using the vertical electric field of the transmitted VLF wave as a phase reference.

Interpretations of the conductivity variations are based on the amplitude ratios of these two components.

Barringer E-phase. In this case the electric field vector of the electromagnetic wave is utilized.

When a conductor is present, there is a small horizontal component of the secondary electric field lying in the direction of propagation from the transmitter; this undergoes field strength variations as the conductivity varies and is out-of-phase with the primary vertical electric component. Over resistive ground the horizontal out-phase component of the secondary E-component (see Figure 7.6) is relatively large whereas over conductive ground it is relatively small. Interpretation is made on this basis. This system is one of the few to make use of phase relationships of the electric vector of the EM wave.

McPhar KEM. Measurements are made of the ratio of horizontal and vertical compnents of the resultant magnetic field and displayed automatically as a dip angle.

Geonics Em 18. Makes use of the relative phase angle of the primary and secondary magnetic fields.

Scintrex Deltair. The gradients of phase and amplitude of the horizontal magnetic component are used.

Induced Pulse Transient (INPUT) System

This is an airborne system of considerable interest. The transmitting loop is essentially horizontal and strung around a large aircraft; the receiver is a towed bird. As the title of the system suggests the method depends on inducing eddy currents in subsurface conductors by means of a short, strong electromagnetic pulse. The decay characteristics of the eddy currents are then analyzed during the "off" period of the transmitter.

The primary pulse is essentially a half sine wave, approximately 1.5 millisecond long, and the quiet period is approximately 2 milliseconds.

If there are no conducting zones then there are no decaying eddy currents in the off period and the primary current (and therefore the magnetic field) and the current induced in the receiver by the primary field will appear as in Figure 7.29(A) and (B). If there are anomalous subsurface conductors, the eddy currents induced by the primary field have a finite decay time and the resultant fields detected by the receiver will appear as shown in Figure 7.29(C). The shape of the decay curve for the eddy currents will give an indication of the subsurface conductivity.

Clearly, it would be highly desirable to trace the complete curve but since this is not possible the response is sampled at a number of intervals, usually of 100 microseconds duration, during the off period.

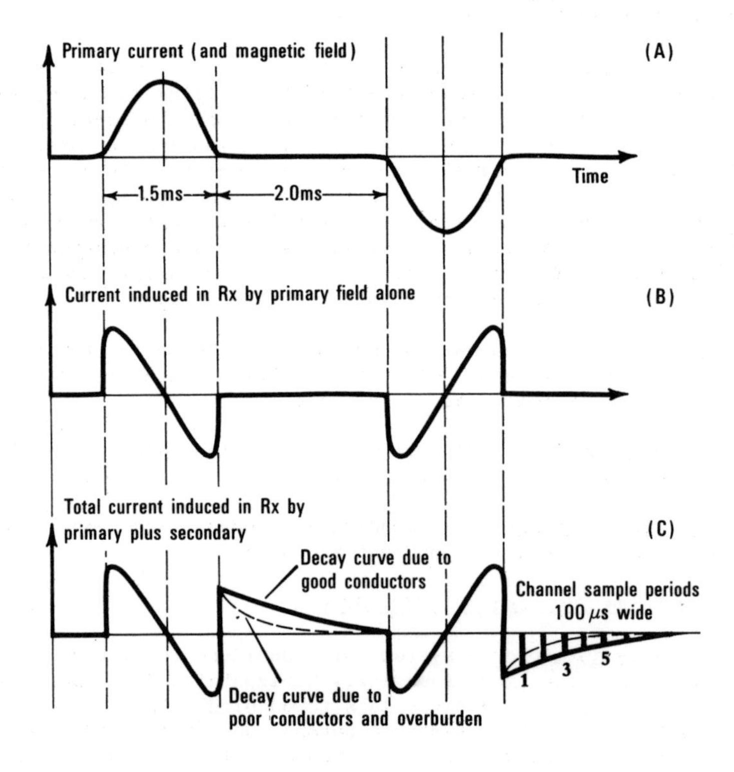

Figure 7.29. Principles of the Induced Pulse Transient (INPUT) system.

If, over the sample period of time, the area under the curve is integrated and the integral displayed for each channel, then the profile would appear as shown in Figure 7.30. As can be seen, good conductor anomalies are likely to show up as distortions in all channels whereas overburden anomalies or poor conductors will show up only in the early channels.

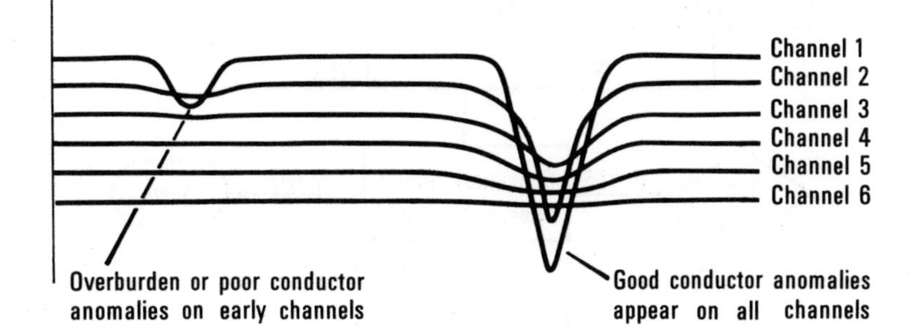

Figure 7.30. Showing how different anomalies appear on INPUT system. Channel numbers correspond with those in Figure 7.29.

Turair

This is essentially a Turam system but with the phase gradients and intensity gradients being measured by airborne receivers. A ground loop is laid out in the same manner as for Turam.

GROUND FOLLOW UP

Airborne electromagnetic (AEM) surveys are only run if there is a reasonable probability of discovering a base metal deposit of particular electrical and geometrical properties, in the range of matching of the system chosen. Ground follow up is necessary to prove, or discard, the AEM anomalies. The flow chart in Figure 7.31 illustrates a typical, but by no means unique, procedure.

The first step is to locate the AEM anomaly precisely on the ground. This is usually done by using a ground EM system chosen in accordance with conductor depth and conductivity thickness estimated from the AEM anomaly, and the type of ground cover in the area. This might be followed by an IP survey to separate the ionic from electronic conductors or occasionally, if there is reason to suspect from the beginning that the AEM anomaly may be

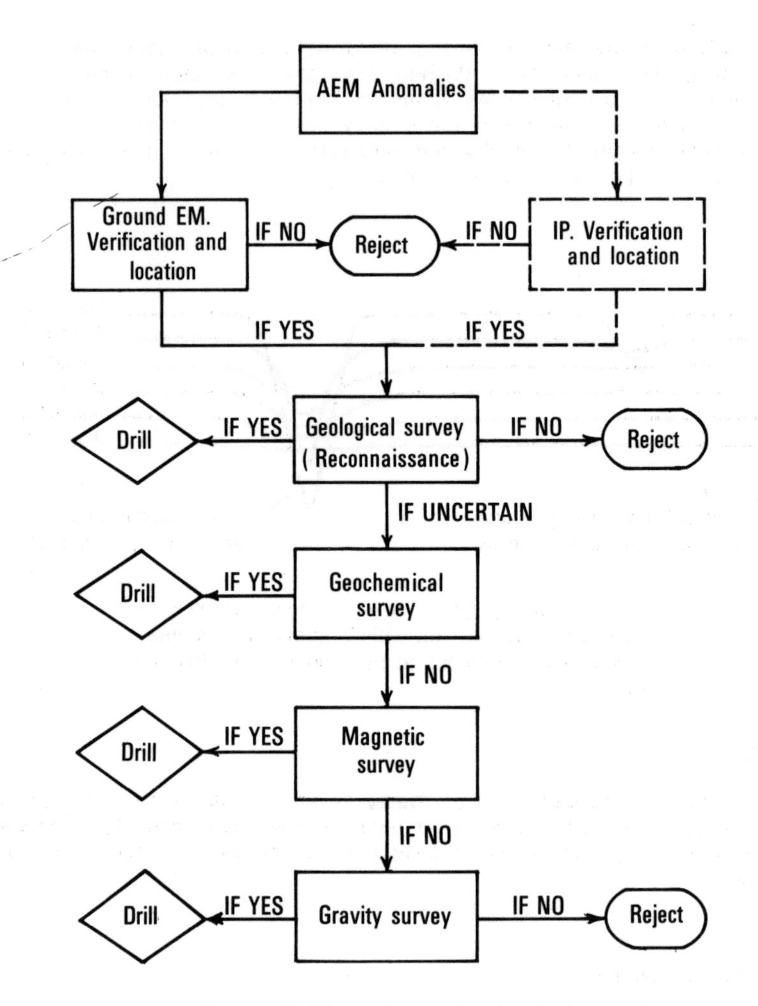

Figure 7.31. A typical, but not unique, methodical approach to ground follow up of airborne electromagnetic anomalies.

due to overburden conduction, the IP survey may be carried out immediately. However, because of the relative expense of the IP survey, this procedure is often bypassed.

If the ground survey verifies the AEM anomaly, a reconnaissance geological survey is the next most likely step. If there is still uncertainty about the value of the anomaly and there are residual soils present, a geochemical reconnaissance may be in order. Should there still be some uncertainty a magnetic survey would indicate if the EM anomaly is associated with bodies containing iron sulphides or oxides; however, since many base metal deposits do not have significant quantities of such minerals, a negative result is not conclusive. The final geophysical test would therefore be a gravity survey since base metal bodies have density contrasts of 1 or 2 gm cm^{-3}; the gravity survey may also give information on the total mass anomaly (see Chapter 5).

SUGGESTIONS FOR FURTHER READING

Bosschart, R.A., 1970. Ground electromagnetic methods, pp. 67-80, in Mining and Groundwater Geophysics, ed. L.W. Morley, Econ. Geol. Report No. 26, Geological Surv. Can., Ottawa.

Bosschart, R.A. and Seigel, H.O., 1966. Some aspects of the Turam electromagnetic method, Trans. C.I.M.M., 69, 156-161.

Fraser, D.C., 1969. Contouring of VLF-EM data, Geophysics, 34, pp. 958-67.

Grant, F.S. and West, G.F., 1965. Interpretation theory in applied geophysics, McGraw-Hill, New York.

Hood, P., 1979. Mineral exploration trends and developments in 1978, Canadian Min. J., 100, 28-69.

Keller, G.V. and Frischhnecht, F.C., 1966. Electrical methods in geophysical prospecting. Pergamon, London.

Ward, S.H., 1967. The electromagnetic method, pp. 224-372, in Mining Geophysics Vol. II, eds. D.A. Hansen, R.E. MacDougall, G.R. Rogers, J.S. Sumner and S.H. Ward, Society of Exploration Geophysicists, Tulsa, Okla.

8

Seismic methods

INTRODUCTION

When the earth is disturbed at either an internal or external point by an earthquake or by hitting it with a feather, the disturbance is transmitted to other points by elastic waves; obviously the low energy input disturbance will not be noticed very far from the initiation point. The term "elastic waves" arises because the properties of the waves depend critically upon the elastic constants of the medium through which they travel.

DEFINITIONS OF ELASTIC CONSTANTS

Starting with Hook`s basic law that strain is proportional to stress, we can define a number of elastic constants. When a uniform body is subjected to a force in one direction, Figure 8.1(A), it will, if the force is compressive, become shorter in the direction of the force and longer in the directions orthogonal to the force. If the force is expressed as a pressure, or stress, Youngs modulus, Y, is defined as

$$Y = \frac{\text{uniaxial stress}}{\text{strain parallel to stress}}$$

$$= \frac{\text{applied load/unit area of cross section}}{\text{fractional increase in length}} = \frac{(F/A)}{\Delta\ell/\ell} \quad (8.1)$$

If instead of a uniaxial stress we have a hydrostatic stress, that is the force per unit area is the same in any three mutually orthogonal directions, the body will undergo a change in volume (or bulk), Figure 8.1(B). The bulk modulus of elasticity is defined by

$$k = \frac{\text{volume stress}}{\text{volume strain}}$$

$$= \frac{\text{hydrostatic force/unit area}}{\text{fractional change in volume}} = \frac{(F/A)}{\Delta V/V} \quad (8.2)$$

Since k is a measure of compressibility, its inverse is often called the incompressibility.

It is also possible to apply a pair of forces such that the body is distorted without a change of volume; that is, it undergoes a shear, Figure 8.1(C). The shear modulus of elasticity (or rigidity modulus), n, is defined by

$$n = \frac{\text{shear stress}}{\text{shear strain}} \tag{8.3}$$

$$= \frac{\text{tangential force/unit area}}{\text{angular deformation}} = \frac{(F/A)}{\theta}$$

Alternatively, we can show that a shear of θ can be regarded as a fractional elongation θ/2 along one diagonal and a fractional contraction θ/2 along the orthogonal diagonal due to a pair of forces at π/4 to the original one.

From the definition of Youngs modulus, it is clear that we have ignored a change in dimensions in directions orthogonal to the direction of the uniaxial stress. This transverse strain can

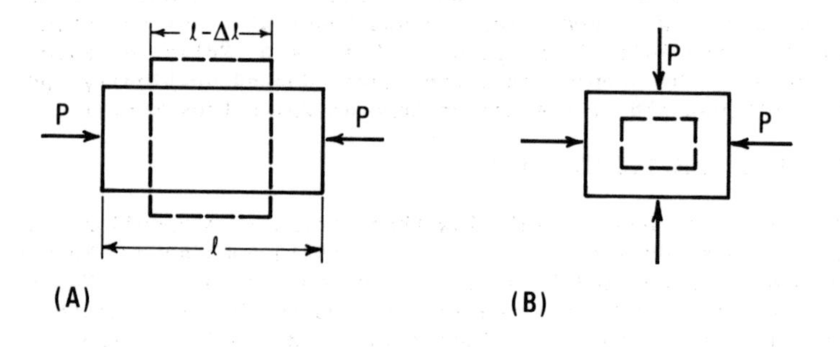

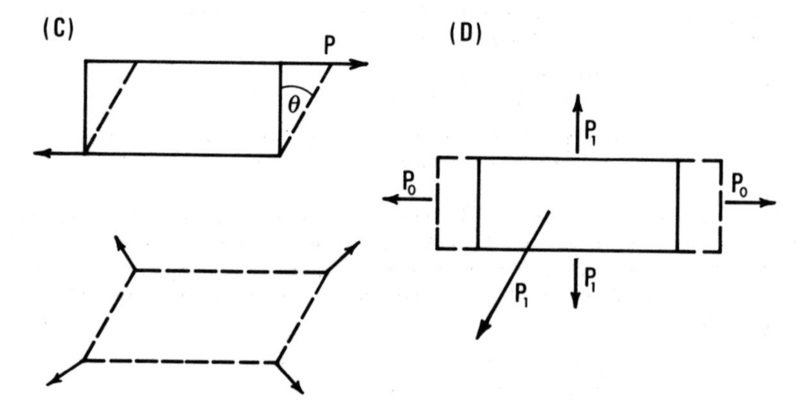

Figure 8.1. Illustration for deriving relationship between stress and strain for (A) Uniaxial stress, (B) Hydrostatic stress, (C) Shear stress, (D) System of non hydrostatic stresses.

be taken into account in two ways. We can define Poisson's ratio, σ, by

$$\sigma = \frac{\text{transverse strain}}{\text{longitudinal strain}}$$

$$= \frac{\text{fractional change in transverse dimensions}}{\text{fractional change in longitudinal dimensions}}$$

$$= \frac{\Delta W/W}{\Delta \ell/\ell} \tag{8.4}$$

Alternatively, if a system of mutually orthogonal forces is applied such that there is elongation, say, in the direction of the principal force P_o and the subsidiary forces P_1 are chosen so as to allow no net change in the dimensions of the body in the directions at right angles to the direction of P_o, Figure 8.1(D), we can define the axial modulus of elasticity, by

$$\chi = \frac{\text{principal stress}}{\text{strain parallel to principal stress}} = \frac{(F/A)p}{\Delta \ell/\ell} \tag{8.5}$$

The verbal definition of χ does not look very different from that of Youngs modulus, but it must be remembered that χ is defined on the basis of a system of forces; in fact P_1 is given by

$$P_1 = \frac{\sigma P}{1-\sigma} \tag{8.6}$$

In seismology, the axial modulus and the shear modulus are of great importance since they control the velocity of compressional and shear waves.

From the way in which we have defined the elastic constants it is clear that they are not independent. They are connected by the following relations.

$$Y = 3k(1-2\sigma) = 2n(1+\sigma) = \frac{\chi(1+\sigma)(1-2\sigma)}{1-\sigma} \tag{8.7}$$

TYPES OF SEISMIC WAVES

There are three principal types of seismic waves defined on the basis of particle motion.

Compressional (P) Waves

The direction of particle motion is the same as the direction of advance of the wave, Figure 8.2(A). This is the fastest travelling wave with a velocity V_p given by

$$V_p = \left(\frac{\chi}{\rho}\right)^{\frac{1}{2}} = \left(\frac{k + 4n/3}{\rho}\right)^{\frac{1}{2}} \qquad (8.8)$$

where ρ is the density of the medium.

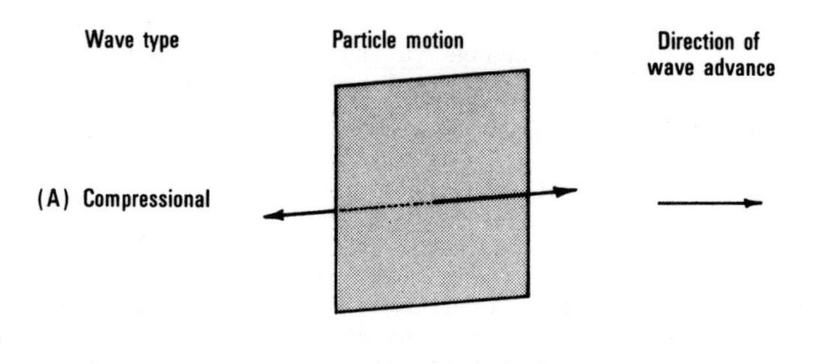

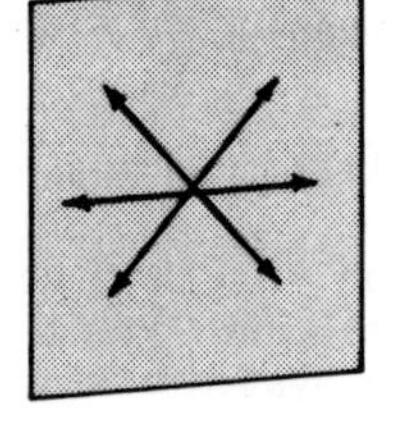

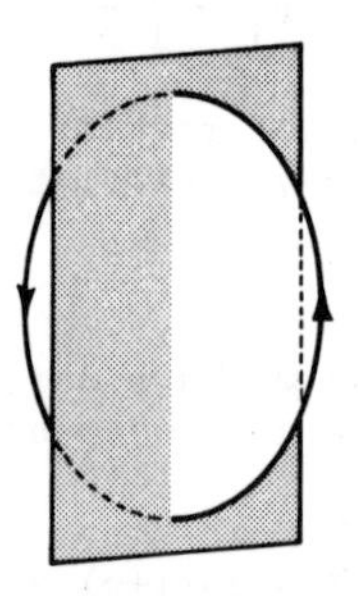

Figure 8.2. Illustrating the relationship between particle motion and direction of travel for the three basic wave types. (A) Compressional, (B) Shear, may be polarized in which case if the polarization is horizontal the wave may be confined to surface layers, and called a Love wave; (C) Rayleigh, major axis with magnitude of major axis decreasing rapidly with depth, i.e. wave is confined to surface layers.

Shear (S) Waves

The directions of particle motion are orthogonal to the direction of advance of the wave Figure 8.2(B). This is the second fastest travelling wave with a velocity V_s given by

$$V_s = (n/\rho)^{\frac{1}{2}} \qquad (8.9)$$

Shear waves which are polarized vertically or horizontally frequently occur and are denoted as SV and SH waves respectively. In earthquake seismology a special case of SH wave is the Love wave (velocity V_L) which is confined to the upper layers of the earth.

Rayleigh Waves

The direction of particle motion is a retrograde ellipse with the major axis vertical and the minor axis being in the direction of advance of the wave, Figure 8.2(C). This has the lowest velocity (V_R) of the waves discussed so far.

The expressions for V_p and V_s do not contain a frequency term and are therefore not subject to dispersion. V_L and V_R are complicated expressions containing frequency terms so that they undergo dispersion. In general

$$V_p > V_s > V_L > V_R$$

There is a great multiplicity of terms for the same waves in different aspects of geophysics.

In earthquake seismology the term "body waves" is used to cover those compressional (Longitudinal, P, Primary or Push) and shear (Shake, Secondary or S) waves that travel deep into the earth`s interior. The term "surface waves" is used to cover Love (Lq, LQ, or G) and Rayleigh waves (R or LR). Other terms, such as "Torsional ringing" and "Spheroidal ringing" are probably best regarded as special cases of Love and Rayleigh wave trains respectively.

In exploration seismology, the emphasis is on the travel time of the compressional wave from point to point. Shear waves have rarely been used, amplitude analysis of the compressional waves is occasionally useful, and Rayleigh waves are only important when they become a nuisance frequently known as "ground roll".

SEISMIC DETECTORS

Seismic disturbances are detected by geophones or seismometers, the latter term usually referring to geophones used in earthquake seismology. The principle behind the geophone is that any relative motion between a magnetic field and an electronic conductor will produce a voltage across the ends of the conductor, the amplitude depending, amongst other things, on the rate at which magnetic flux cuts the conductor. It is not really important which of the conductor or magnet is kept fixed provided one is free to move relative to the other, Figure 8.3.

The voltage is amplified and recorded by a variety of methods. For a completely flexible geophone the voltage would follow the motion of the earth at the point where it is situated; in practice, the finite strength of the spring and "feedback" effects introduce distortion to the wave form (that is the output voltage is not a faithful reflection of the ground motion) even before electronic amplification takes place, which may introduce further distortion. It is often difficult to allow for these problems in exploration methods, where frequencies of 10 to 100 Hz are used to maximize resolution, so that travel-time analysis, rather than wave-form analysis, is the normal way to obtain useful information. However, many of the techniques of data analysis

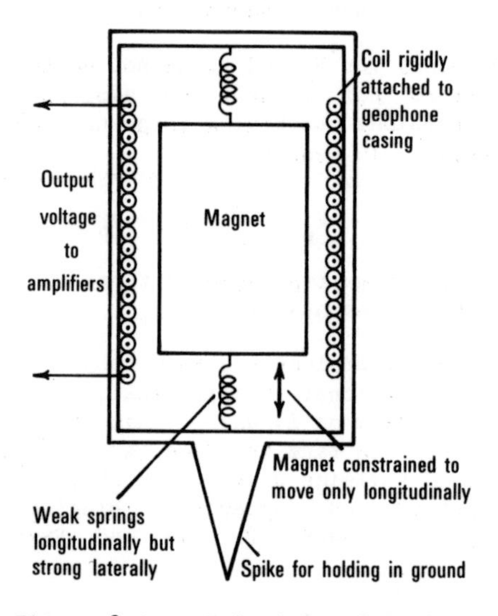

Figure 8.3. Principle of operation of seismic detectors (geophones or seismometers). When the case moves with the ground, the inertia of the magnet causes relative motion and the moving magnetic field induces a voltage in the coil. Usually for exploration work the vertical component of the disturbance is detected; occasionally, and always for earthquake seismology, three component geophones are used.

used in exploration methods can be used in earthquake seismology, and vice versa. This may be obvious because the spectrum of disturbances covers a wide range with ever increasing overlap, Figure 8.4.

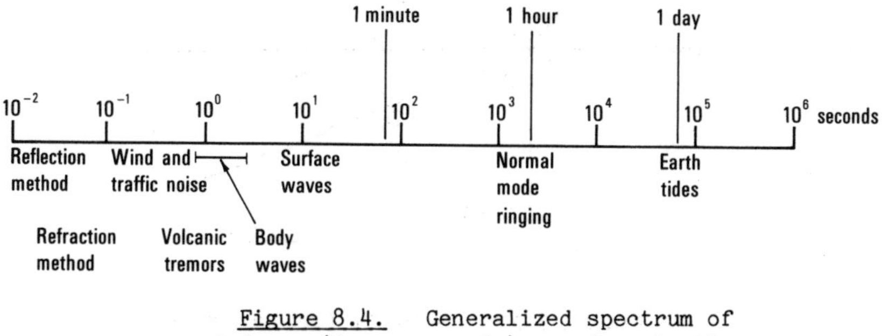

Figure 8.4. Generalized spectrum of periods (in seconds) of elastic disturbances to the earth.

REFLECTION AND REFRACTION

Reflection and refraction of seismic waves occurs when there is a difference in acoustic impedances between two media, the acoustic impedance being defined as the product of density (ρ) and velocity (V). Therefore, when we refer to discontinuities, we really mean a discontinuity in acoustic impedance, ρV, although usually the density variation is so small compared with velocity changes we often automatically think in terms of velocity discontinuities.

The reason why little character analysis of exploration seismograms is undertaken is that the wave forms are too complex even if the detector, or geophone, which traces the motion of the disturbed earth at a particular point is completely distortion free. The complexity arises because the elastic constants are not independent; in general, a compressional wave reflected or refracted at a discontinuity will give rise to reflected and refracted shear waves as well as compressional waves, and the shear waves, which are always present even if their information content is not utilized, give reflected and refracted compressional waves as well as more shear waves. Since seismic surveys are carried out in areas where there are many discontinuities, it is obvious how the complexity arises.

The Huyghens principle established for light waves, can be

used to determine the laws of refraction and reflection of seismic waves. Consider a portion of a plane compressional (P) wave incident upon a velocity discontinuity, Figure 8.5.

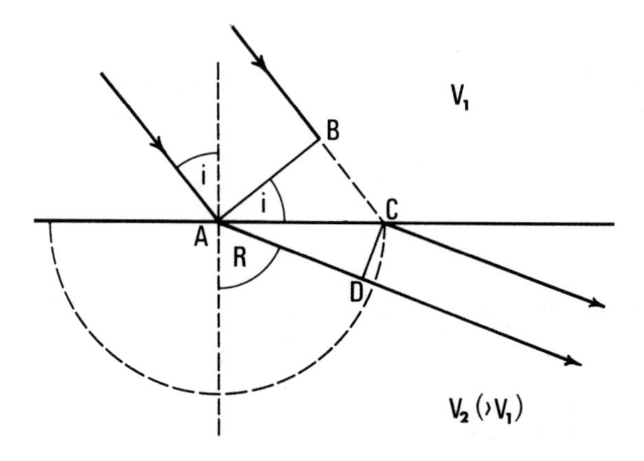

Figure 8.5. Geometry for refracted wave front at a seismic discontinuity.

When the disturbance reaches point A on the discontinuity it will commence being transmitted at V_2 in medium 2. The time required for the disturbance to travel from B to C in medium 1 is simply $t = BC/V_1$; in this time the disturbance in medium 2 travels a distance greater than BC, since $V_2>V_1$, to point D so that $AD = BC\ (V_2/V_1)$. Intuitively it can be seen that CD will form the refracted wave front. From the geometry of Figure 8.5.

$$\frac{\sin i}{\sin R} = \frac{BC/AC}{AD/AC} = \frac{BC}{AD} = \frac{V_1}{V_2} \tag{8.10}$$

In this example we have used V_1 and V_2 to denote compressional wave velocities; however, if we now use V_2^S to denote the shear wave velocity in medium 2, exactly the same relationship as 8.10 will hold with the substitution of V_2^S for V_2 and R_s for R. When $R = \pi/2$, $\sin i = V_1/V_2$ and the angle of incidence is called the critical angle i_c, or $i_{n,n+1}$ when several media are involved.

By a similar process we can derive the laws of reflection; for example, using an incident S-wave and a reflected P-wave, we get

$$\frac{\sin i_s}{\sin r_p} = \frac{V_s}{V_p} \tag{8.11}$$

where i_s is the angle of incidence of the S-wave, r_p the angle of reflection of the P-wave, V_s and V_p are the shear and compressional wave velocities respectively. For reflected shear waves, the angle of reflection equals the angle of incidence.

The amplitudes of the reflected and refracted (transmitted) waves can be predicted from complex equations which reduce to a very simple form if the incident wave is normal, or within about 15° of the normal, to the discontinuity. In this case if A_i is the amplitude of the incident energy, and A_r and A_t the reflected and transmitted amplitudes

$$\frac{A_r}{A_i} = \frac{\rho_2 V_2 - \rho_1 V_1}{\rho_2 V_2 + \rho_1 V_1}$$

and

$$\frac{A_t}{A_i} = 1 - \frac{A_r}{A_i}$$

where A_r/A_i and A_t/A_i are often called the reflection and transmission coefficients. Unfortunately there is some confusion here since the corresponding fractions of reflected and transmitted energy (which varies as the square of the amplitude) are also called the reflection and refraction coefficients.

WAVE FRONTS AND RAYS

It is useful to employ a concept of rays, or lines of elastic disturbance in the direction of advance of the waves. Many problems become very easy to understand and simple to analyse but it must be remembered that the concept has no physical reality whatsoever. Since it is a convenience measure only, albeit a quite powerful one, when results become difficult to understand it might be wise to resort to full wave front theory in case a rare situation has arisen where simple ray theory cannot be applied.

The reasoning behind the use of rays, and the reason why it is a powerful simplifying tool, may be seen from Figure 8.6.

Energy first reaches the V_2 medium at point A. From this point on, the disturbance is transmitted along the interface at a velocity V_2; the energy in this disturbance is referred to as the head wave energy. Since each point on the interface acts as a source of energy it transmits energy back into the V_1 medium as well as into the V_2 medium. However, until point B is reached, the energy first arrives at points in the V_1 medium and on the interface via the V_1 medium alone. After point B the energy first arrives at points in the V_1 medium via the retransmission of head wave energy from the interface. Applying the Huyghens principle, that each point on the transmitted disturbance becomes a source for a spherical disturbance, by constructing successive wave front diagrams it is possible to show that the head wave energy in the V_1 medium forms a plane wave. The curve connecting the points in the V_1 medium at which energy arrives simultaneously via the direct and indirect route is called the equal time locus. The equal time locus intersects the surface at point C whose distance from the source is called the "critical distance"; in recent years there have been attempts to replace this term with "crossover distance" and use "critical distance" to refer to the particular value of the offset distance (see later section) in the reflection method when the angle of reflection equals the critical angle.

If from point C we draw a line perpendicular to the plane wave front, the angle this line makes with the vertical is the critical angle i_{12}; if we now draw a line from S to the interface at an angle i_{12} to the vertical, and draw in the horizontal line connecting them along the interface, this line SDEC is called the <u>critical ray</u>. It is simply a convenient geometrical

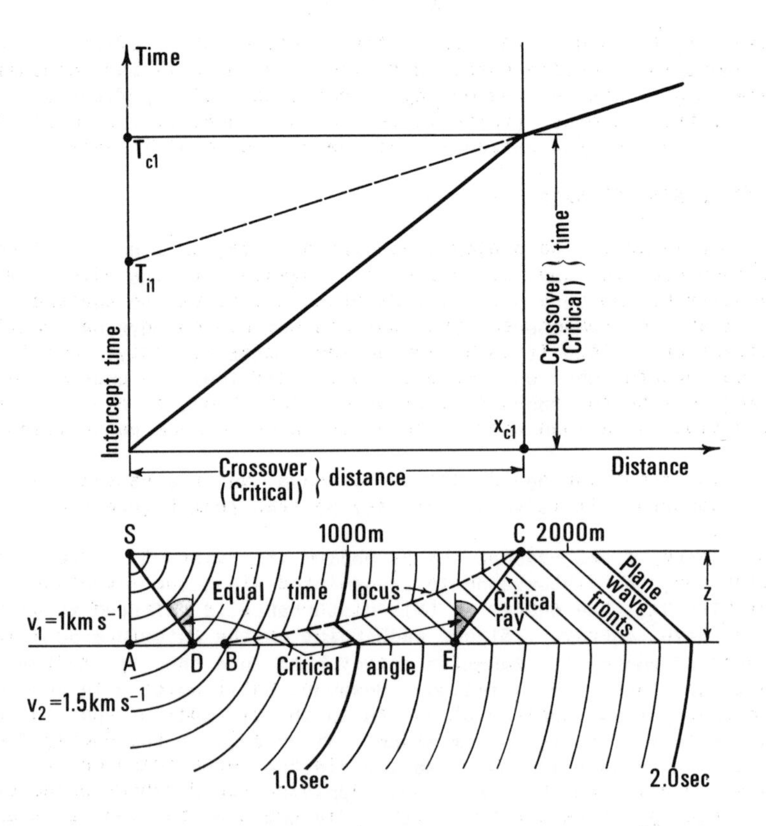

<u>Figure 8.6.</u> Illustrating why ray diagrams may be substituted for wave front diagrams and origin of concepts of critical ray, and its relationship to the distance-time graph in refraction methods.

representation of the passage of energy via the V_2 medium which happens to arrive at C at the same time as direct energy from S via the route SC. The geometrical manipulations applied to ray theory are simple to follow and are applicable to all cases in this book; we shall therefore refer to "rays" from now on rather than to wave fronts.

Similar constructions and arguments can be made for multi-layer cases, with parallel or dipping interface. In all cases it is usually assumed that $V_{n+1}>V_n$.

The first packet of energy to arrive at a detector, by any route, is called the first arrival; if the energy is from a refracted wave, e.g. from any point beyond C in Figure 8.6, it is the first refracted arrival and if from a point before C it is the direct arrival. As will be seen later, no reflected energy can be first arrival energy.

The distance AB depends upon the velocity contrast, being large when the contrast is small. The fact that the equal time locus cannot intersect the interface vertically below the source S suggests that in multilayer cases there may be some occasions when the velocity contrast and thickness of the (n+1)th layer (which lies between the nth and (n+1)th interfaces) are such that head wave energy from the (n+1)th interface reaches the surface before head wave energy from the nth interface. In such cases there are no first arrivals from the nth interface and more sophisticated means must be used to detect the "hidden" V_{n+1} layer; this problem is quite different from the hidden layer problem when $V_{n+1}<V_n$.

REFRACTION METHODS

Two Layer, One Discontinuity Parallel to Surface

Figure 8.7(A) shows the critical ray for a two layer (one velocity discontinuity) case where the interface is parallel to the surface. The time T_{c1} for the disturbance to travel from S to C by the direct route is x_{c1}/V_1 and by the indirect route is $t_{SA} + t_{AB} + t_{BC}$. From symmetry, $t_{SA} = t_{BC}$ so that

$$T_{c1} = \frac{x_{c1}}{V_1} = \frac{2SA}{V_1} + \frac{AB}{V_2}$$

$$= \frac{2z}{V_1 \cos i_{12}} + \frac{x_{c1} - 2z \tan i_{12}}{V_2} \qquad (8.12)$$

From (8.10) $\sin i_{12} = V_1/V_2$ and all other angular functions of i_{12} can therefore be obtained in terms of V_1 and V_2. Substitution of these functions in (8.12) and subsequent simplification gives

$$z = \frac{x_{c1}}{2}\left(\frac{V_2-V_1}{V_2+V_1}\right)^{\frac{1}{2}} = \frac{x_{c1}}{2}\left(\frac{m_1-m_2}{m_1+m_2}\right)^{\frac{1}{2}} \quad (8.13)$$

where $m_1 = 1/V_1$ and $m_2 = 1/V_2$ and are the slopes of the segments of the distance time graph shown in Figure 8.7(A). That m_1 is the slope of the first segment is obvious; that m_2 is the slope of the second segment may seem strange since it implies that the disturbance in medium 1 is travelling at velocity V_2 on the surface. That this is so may be seen from Figure 8.7(B). Consider the wave front DE. When the disturbance in medium 1 has reached the surface at D, it has only reached E internally. It therefore has to travel distance EF in, say, time dt which is given by EF/V_1. But $EF = dx \sin i_{12}$ so that

$$dt = \frac{dx \sin i_{12}}{V_1}.$$

or

$$\frac{dx}{dt} = \frac{V_1}{\sin i_{12}}$$

but since we are considering the critical angle i_{12}, $\sin i_{12} = V_1/V_2$ from (8.10), therefore

$$\frac{dx}{dt} = V_2 \quad (8.14)$$

but dx/dt is an expression of the velocity of the disturbance on the surface so that the second segment of the distance - time graph, Figure 8.7(A), gives V_2, and therefore m_2.

An alternative method of estimating z, is to use the intercept time T_{i1}. From Figure 8.7(A), it can be seen that

$$\frac{x_{c1}}{V_2} = T_{c1} - T_{i1} \quad (8.15)$$

Using (8.12) and (8.15) and rearranging we get

$$z = T_{i1}/2(m_1^2 - m_2^2)^{\frac{1}{2}} \quad (8.16)$$

Two Layer - Dipping Discontinuity

Figure 8.8(A) Shows the critical ray for the case where the velocity discontinuity dips at angle α relative to the surface and the detectors are placed downdip from the sources. As before, the downdip travel time, T_{c1d}, via the direct route SC is the same as via the indirect route SABC, and is given by

$$T_{c1d} = \frac{Z_d \sec i_{12}}{V_1} + \frac{(Z_d + x_{c1d} \sin\ )\sec i_{12}}{V_1}$$

$$+ \frac{x_{c1d} \cos \alpha - (Z_d + Z_d + x_{c1d} \sin \alpha)\tan i_{12}}{V_2} \quad (8.17)$$

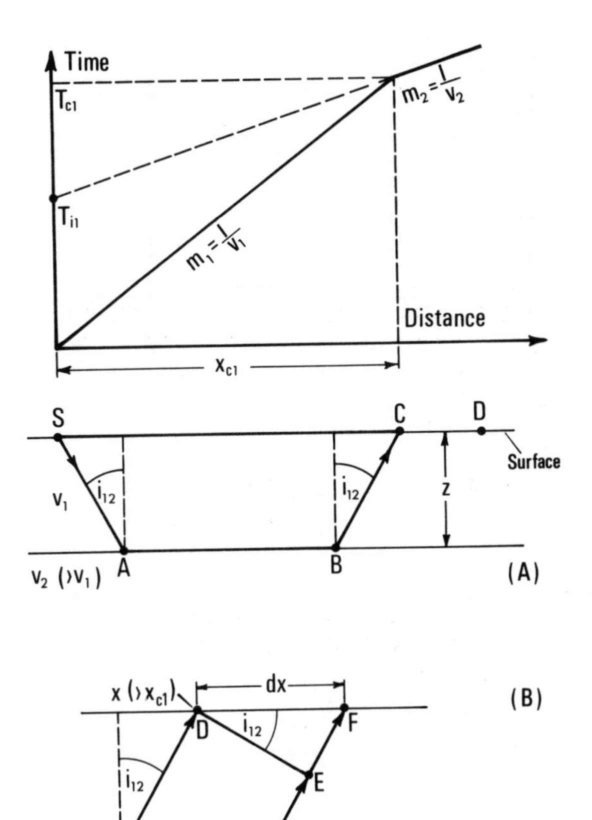

Figure 8.7. Two layer case with the discontinuity parallel to the surface. (A) Critical ray path, distance-time graph and useful quantities derived from the graph (B) Illustrating how disturbance from discontinuity can travel at a velocity of V_2 along the surface of a V_1 layer.

The function of i_{12} can again be expressed in terms of V_1 and V_2, and substitution and appropriate reduction leads to the result

$$T_{c1d} = \frac{x_{c1d}}{V_{2d}} + 2z_d\left(\frac{1}{V_1^2} - \frac{1}{V_2^2}\right)^{\frac{1}{2}} \qquad (8.18)$$

where V_1 and V_2 are the true velocities in the first and second media and V_{2d} is the apparent velocity in the second medium when the detectors are downdip of the source. The intercept time is

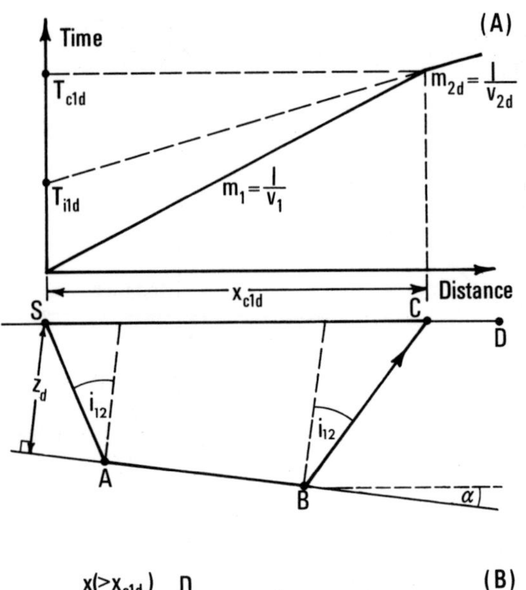

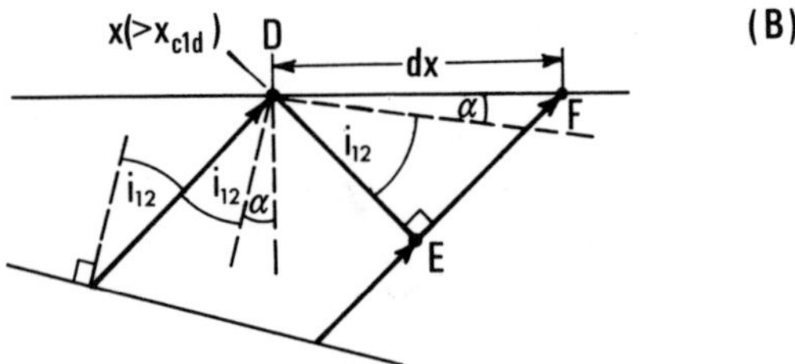

<u>Figure 8.8.</u> Two layer case with a dipping discontinuity. (A) Critical ray path, distance - time graph and useful quantities derived from the graph. (B) Showing how disturbance on surface travels at a velocity which depends upon α as well as V_1 and V_2.

again a useful alternative to use and from Figure 8.8(A) it can be seen that

$$T_{i1d} = T_{c1d} - \frac{x_{c1d}}{V_{2d}} = 2z_d\left(\frac{1}{V_1{}^2} - \frac{1}{V_2{}^2}\right)^{\frac{1}{2}} \tag{8.19}$$

By similar arguments we can find that when the detectors are up dip of the source

$$T_{c1u} = \frac{x_{c1u}}{V_{2u}} + 2z_u \left(\frac{1}{V_1{}^2} - \frac{1}{V_2{}^2}\right)^{\frac{1}{2}} \tag{8.20}$$

$$T_{i1u} = 2z_u \left(\frac{1}{V_1^2} - \frac{1}{V_2^2}\right)^{\frac{1}{2}} \tag{8.21}$$

From the pairs of equations, 8.18, 8.20 and 8.19, 8.21, it seems at first sight that we have three unknowns and only two equations. T_{c1d}, x_{c1d}, T_{i1d}, T_{c1u}, x_{c1u}, T_{i1u}, V_{2d}, V_{2u} and V_1 can be found directly from the distance time graphs. We wish to determine z_d and z_u but do not know V_2 (nor the implied dip angle α). Fortunately, as may be seen from Figure 8.8(B), the apparent velocities in the second layer can be determined from the distance time graph and the apparent velocities depend on the true velocities in media 1 and 2 and on the dip angle. Consider the plane wave front. When the disturbance in the medium of velocity V_1 reaches the surface at A it has only reached B internally. The time for the disturbance to travel BC at velocity V_1, is given by

$$dt = BC/V_1 = dx \sin (i_{12} + \alpha)/V_1$$

or

$$\frac{dx}{dt} = \frac{V_1}{\sin(i_{12} + \alpha)} = V_{2d} \tag{8.22}$$

Where V_{2d} is the apparent velocity in the second medium and can be found from the slope $m_{2d} = 1/V_{2d}$ of the second line segment of the down dip shot. Similarly

$V_{2u} = V_1/si$

$$V_{2u} = \frac{V_1}{\sin(i_{12} - \alpha)} \tag{8.23}$$

Since V_1 can be measured from the first segment of the distance-time curves, (8.22) and (8.23) contain only two unknowns, i_{12} and α, and they can be solved to give

$$\alpha = \frac{1}{2} [\sin^{-1}(\frac{V_1}{V_{2d}}) - \sin^{-1}(\frac{V_1}{V_{2u}})] \tag{8.24}$$

and

$$i_{12} = \frac{1}{2} [\sin^{-1}(\frac{V_1}{V_{2u}}) + \sin^{-1}(\frac{V_1}{V_{2d}})] \tag{8.25}$$

Since we now know i_{12} and we already know V_1, we can find V_2 from (8.10). Equations (8.18) - (8.21) can therefore be solved for the remaining two unknowns.

Three Layers - Discontinuity Parallel to Surface

Figure 8.9 shows the critical ray for refraction from the second discontinuity of a three layer system where the discontinuities are parallel to the surface.

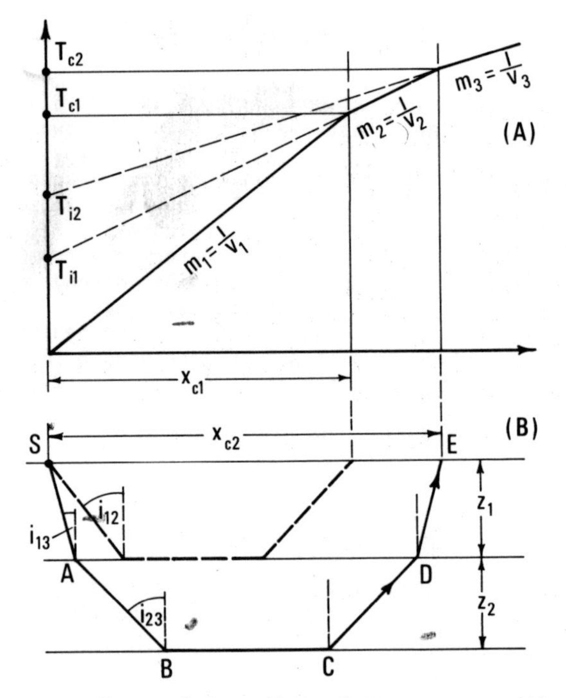

Figure 8.9. Three layer case with discontinuities parallel to the surface.

For the critical ray SABCDE shown, the angle of incidence at the first discontinuity is the same as the critical angle for layer 1 immediately on top of layer 3. This follows from (8.10) since $R = i_{23}$ so that

$$\sin i = \frac{V_1}{V_2} \sin i_{23} = \frac{V_1}{V_2} \cdot \frac{V_2}{V_3} = \frac{V_1}{V_3} = \sin^{-1}(i_{13}) \qquad (8.26)$$

At point E, the first energy arrival at the surface changes from that due to the interface 1 headwave to that due to the interface 2 headwave. Beyond E the velocity of the surface disturbance is V_3.

It can be seen from the geometry of Figure 8.9 that the time T_{c2} for the energy to reach E from S via ABCD is given by

$$T_{c2} = \frac{2z_1 \sec i_{13}}{V_1} + \frac{2z_2 \sec i_{23}}{V_2} + \frac{x_{c2} - 2(z_1 \tan i_{13} + z_2 \tan i_{23})}{V_3} \qquad (8.27)$$

All functions of i_{13} and i_{23} can be expressea in terms of V_1, V_2 and V_3. Substitution of these in 8.27 and subsequent simplification leads to

$$T_{c2} = 2z_1 \sqrt{\frac{1}{V_1^2} - \frac{1}{V_3^2}} + 2z_2 \sqrt{\frac{1}{V_2^2} - \frac{1}{V_3^2}} + \frac{x_{c2}}{V_3}$$

$$= 2z_1(m_1^2 - m_3^2)^{\frac{1}{2}} + 2z_2(m_2^2 - m_3^2)^{\frac{1}{2}} + m_3\, x_{c2} \qquad (8.28)$$

Where m_1, m_2 and m_3 are the reciprocals of V_1, V_2 and V_3 and are therefore the slopes of the three line segments of the distance-time graph of Figure 8.9. From this figure it can also be seen that the second intercept time T_{i2} is given by

$$T_{i2} = 2_{z1}(m_1^2 - m_3^2)^{\frac{1}{2}} + 2z_2(m_2^2 - m_3^2)^{\frac{1}{2}} \qquad (8.29)$$

Multilayer Cases

From the form of 8.28 and 8.29 it is easy to see, without working through it, that for the N+1 layer case, with N discontinuities

$$T_{cN} = m_{N=L}\, x_{cN} + 2 \sum_{n=1}^{N} z_n(m_n^2 - m_{N+1}^2)^{\frac{1}{2}} \qquad (8.30)$$

and

$$T_{iN} = \sum_{n=1}^{N} z_n(m_n^2 - m_{N+1}^2)^{\frac{1}{2}} \qquad (8.31)$$

provided $V_n > V_{n-1}$

Similar arguments can be used to derive the appropriate equation for the multilayer case with dipping discontinuities, the form being, of course, much more complex. If detailed analysis is required in refraction records, the reduction formulae are usually manipulated into a form that can be readily iterated on a modern computer.

In some regions there is an almost continuous increase of velocity with depth, which leads to a non linear distance-time curve, Figure 8.10; the slope at any point on the curve gives the velocity in the deepest layer to which the curved critical ray penetrates assuming, of course, that the discontinuities are all parallel with the surface.

The refraction method is used for reconaissance, not

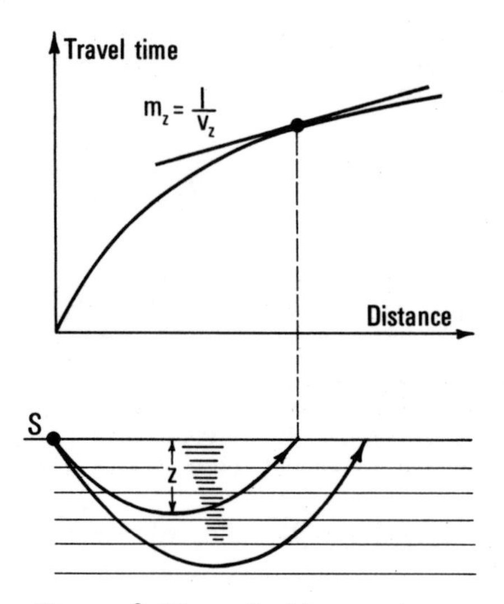

Figure 8.10. Continuous increase of velocity with depth gives curved ray paths and distance-time graph.

detailed, work. If a particular refracting discontinuity is of interest, it can be traced over a long distance fairly easily; however, topography on the discontinuity can rarely be distinguished.

Geophone Layout

A typical geophone layout, or "spread", for profiling is as shown in Figure 8.11(A). The distance between the shot point and the first geophone, and between separate geophones, depends on the depth of the discontinuity and the velocity contrast; but typical distances are S_1A = 1 to 2km and geophone separation = 100 to 200 m with one or two dozen geophones being used; for deep crustal sounding these distances might be multiplied by 10 or as much as 100. To determine possible dip of the discontinuity shots are fired from each end of the spread. To conserve both human and material energy, two geophone spreads are often used, symmetrically placed with respect to the shot points; while this is going on another shot hole, S_3, is being drilled and another spread layed out and the process repeated as often as desired. The profile is usually aligned roughly downdip, the dip direction having been determined by previous experiment. If field conditions are such that the geophone alignment cannot be downdip, separate orthogonal profiles are necessary at intervals.

A technique much used in the early days of prospecting for oil, and still occasionally used for investigation of diapiric structures of any type, is known as "fan shooting", Figure

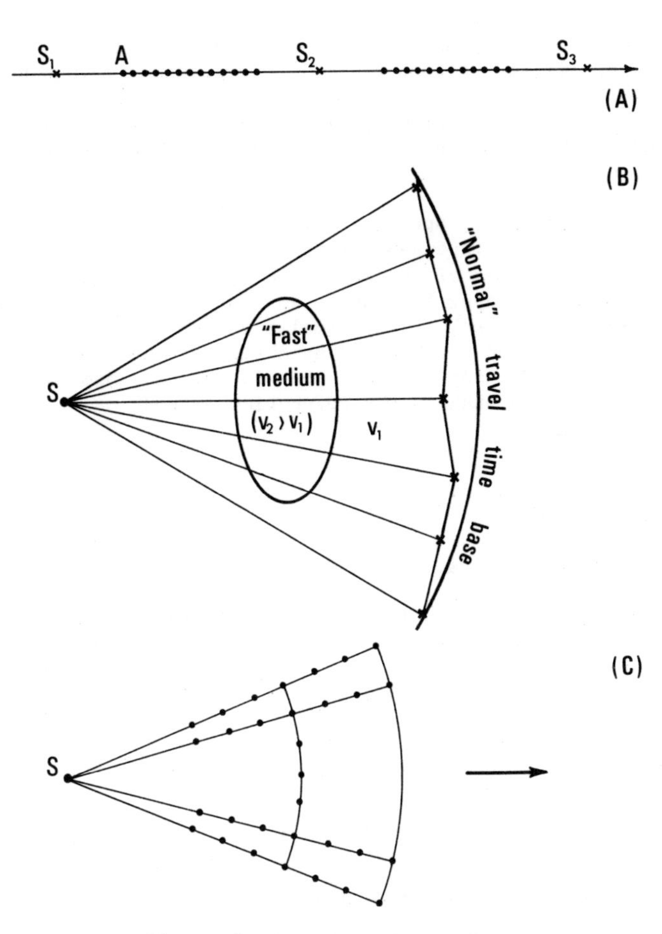

Figure 8.11. Examples of geophone and shot point geometries for refraction methods. (A) In line profiling, (B) Fan shooting for outlining diapiric structures, (C) Arc shooting to give some control on structure in azimuth.

8.11(B). A "normal" travel time for a ray is established by shooting in an area near the structure. Shot points and geophones are then located so that several rays have to pass through the structure while others miss it. If the structure is a high velocity material, such as salt, those rays passing through it will arrive early relative to the "normal" time; the "lead" time for that particular distance is then plotted, the effects of refraction being ignored. Maximum lead time corresponds to maximum thickness of the structure and an idea of its shape and size is thus obtained.

Arc shooting, a modification of this procedure, is used for normal profiling, Figure 8.11(C), the geophone layout being such that there is some control on dip in azimuth without having to arrange a separate group of profiling experiments as would have to be done if the layout of 8.11(A) is used and the profile is not aligned down dip.

Whatever type of spread is used, corrections may be needed for elevation differences between shot point and geophone, and changes in thickness of the weathered layer. Because the velocity in the weathered layer is usually so much less than that in layers 1 km, say, deep, a relatively small change in weathered layer thickness might be interpreted as a significant structural change at depth. These corrections are far more important in the reflection method than in the refraction method.

REFLECTION METHODS

Discontinuity Parallel to Surface

For a ray reflected from the discontinuity parallel to the surface, the travel time T from S to C via B, Figure 8.12(A), is given by

$$T = \frac{SB + BC}{V_1} = \frac{(x^2 + 4z^2)^{\frac{1}{2}}}{V_1} \tag{8.32}$$

The depth to the reflecting horizon is therefore given by

$$z = \frac{1}{2}\sqrt{(V_1 T)^2 - x^2} \tag{8.33}$$

Equation (8.32) can be rearranged to give

$$T^2 = \frac{x^2}{V_1^2} + \frac{4z^2}{V_1^2} \tag{8.34}$$

which is the equation of a hyperbola with the axis of symmetry at x = 0, Figure 8.12(B). Differentiating with respect to x gives

$$\frac{\partial T}{\partial x} = \frac{1}{(1 + \frac{4z^2}{x^2})^{\frac{1}{2}} V_1} \tag{8.35}$$

when $x \to \infty$, $\partial T/\partial x \to 1/V_1$, so that the asymptote to the reflection hyperbola is the slope of the first segment of the refraction distance-time graph.

If $\partial T/\partial x = 1/V_2$, substitution in 8.35 and solving for x gives

$$x = \frac{2zV_1}{(V_2^2 - V_1^2)^{\frac{1}{2}}} = 2z \tan i_{12} \tag{8.36}$$

Thus the slope of the second segment of the refraction travel time-distance curve is tangent to the reflection hyperbola at a point where the shot point and geophone separation is such that the geophone receives the ray which has been reflected at the critical angle; the distance x is now increasingly being referred to as the "critical distance" while the term "crossover distance" is being used in the refraction method for the distance at which both direct and refracted arrivals are observed simultaneously.

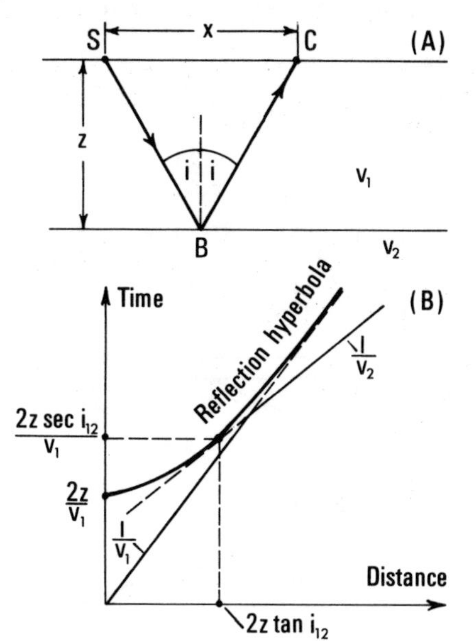

Figure 8.12. (A) Reflected ray diagram for discontinuity parallel to surface; (B) Reflection hyperbola showing relationship with refraction travel time curve.

Dipping Discontinuity

When the discontinuity dips at an angle α the time for the disturbance to travel from S to C via B is, from Figure 8.13(A), $T = (SB + BC)/V_1 = S_1C/V_1$, so that

$$V_1^2T^2 = (S_1C)^2 = (S_1S)^2 + SC^2 - 2S_1S.SC \cos \quad (90 + \alpha)$$

$$= 4z^2 = x^2 + 4xz \sin \alpha \tag{8.37}$$

Differentiating once with respect to x we get

$$2V_1^{\ 2}T\,\frac{\partial T}{\partial x} = 2x + 4z \sin \alpha \qquad (8.38)$$

or putting (8.38) in terms of field measurable quantities

$$\sin \alpha = \frac{V_1^{\ 2}T}{2z} \cdot \frac{\Delta T}{\Delta x} - \frac{x}{2z} \qquad (8.39)$$

where ΔT is the difference in arrival times of reflected disturbances at two geophones separated by a distance Δx.

T, x, ΔT and Δx can be measured in the field so that α and V_1 are unknowns. The problem is therefore insoluble without further inforamtion. Clearly if $\alpha = 0$ then

$$V_1^{\ 2} = \frac{x.\Delta x}{T.\Delta T} \qquad (8.40)$$

To determine dip, we frequently use the shortest return times measured from two adjacent shot points, Figure 8.13(B); from 8.37, if x=0 then obviously $T_1 = 2z_1/V_1$ and $T_2 = 2z_2/V_2$ so that $\sin \alpha = (z_2-z_1)/x = V_1(T_2-T_1)/2x$.

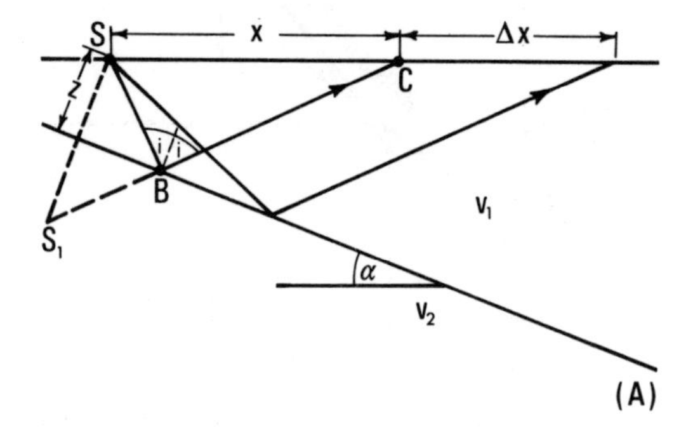

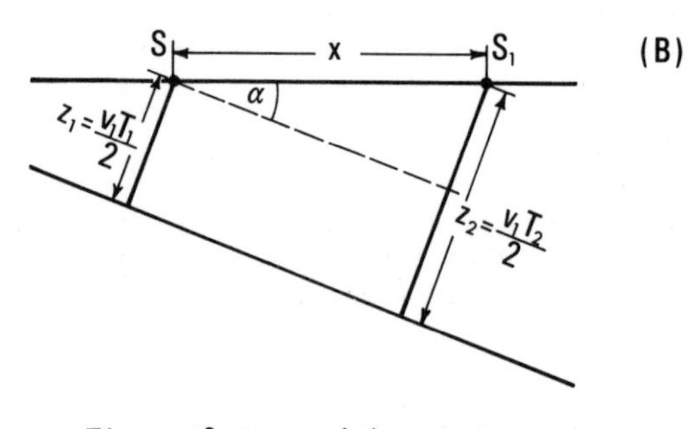

Figure 8.13. (A) Reflected ray diagram for a dipping discontinuity. (B) Method of determining dip angle using shortest return times for two shot points S and S_1.

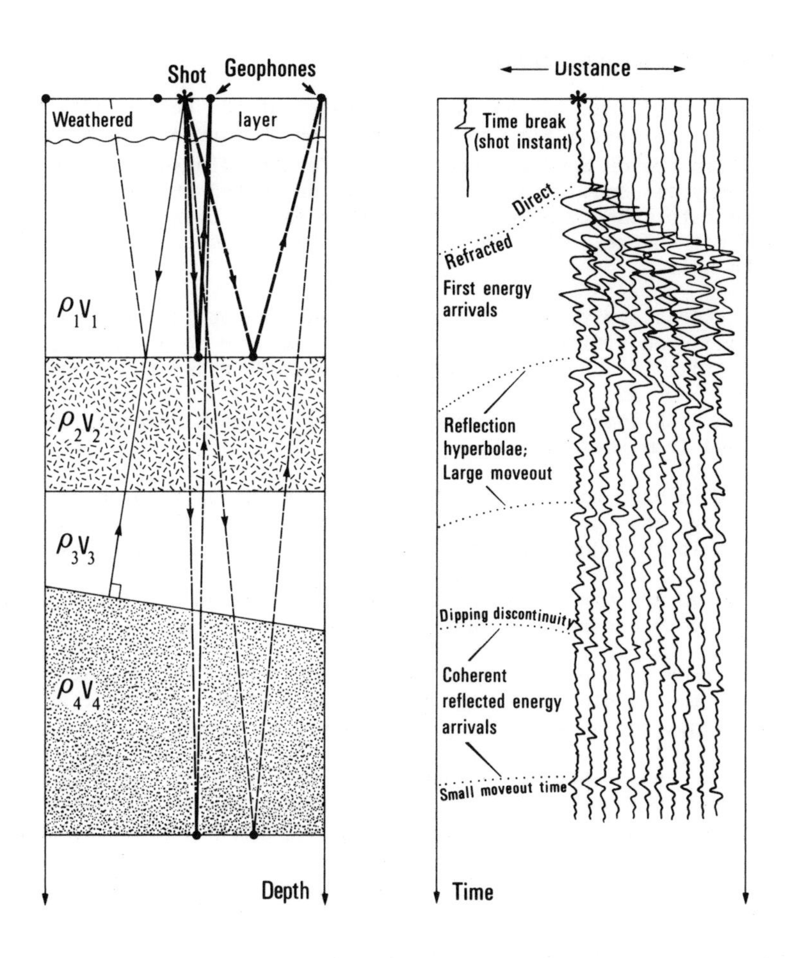

Figure 8.14. Schematic of "wiggle trace" seismogram illustrating coherent energy arriving as A - refracted events (the first energy arrival), and B - reflected events from horizons at various depths.

A typical "wiggle trace" reflection record is illustrated in Figure 8.14. The first part of the record is due to the refracted energy arriving - the first arrivals. Subsequent coherent energy arrivals are due to the reflected event; it can readily be seen that the curvature of the reflection hyperbola becomes less as the reflected time becomes greater, that is as z, Figure 8.12(A), becomes greater, as is expected from equation 8.34. When the hyperbola is asymmetrical with respect to the shot point, dip of the reflecting horizon is indicated.

Methods of Finding Velocity

(a) Well shooting. A charge is detonated at the bottom of a deep hole and the travel time to a geophone on the surface is measured; sometimes two or more geophones at different distances from the hole may be used. A second charge is detonated at a higher level in the hole, the lower part of the hole having been destroyed by the first charge. The process is repeated at several points going up the hole.

Alternatively, a string of geophones can be lowered down the hole and a shot fired at the bottom of the hole or at the top near the casing; in this case there is a risk of losing the geophones.

In both cases the interval velocity and mean velocity can be obtained from the one run.

It is also possible to obtain interval velocities on a fine scale by the use of in-hole continuous velocity devices using electronic transducers as the seismic source (see Chapter 10).

Holes are not always available in the region of interest in which case other methods must be used.

(b) if many measurements have been made then it follows from (8.34) that a plot of T^2 against x^2 for a given horizon will give a straight line of slope $1/V_1^2$ and intercept $4z^2/V_1^2$.

(c) if x_1 and x_2 are the distances of the nearest and farthest geophones from the shot point in a single spread and T_1 and T_2 are the reflection times from a given horizon, then

$$V_1^2 = \frac{x_1^2 - x_2^2}{T_1^2 - T_2^2} \tag{8.41}$$

Equation 8.41 may be rewritten as

$$V_1^2 = \frac{(x_2 + x_1)\ (x_2 - x_1)}{(T_2 + T_1)\ (T_2 - T_1)} = \frac{x_{av}\Delta x}{T_{av}\Delta T} \tag{8.42}$$

where $x_{av} = (x_1 + x_2)/2$, $T_{av} = (T_1 + T_2)/2$, Δx is the geophone separation and ΔT is the arrival time difference between the nearest and farthest geophones; Equation (8.42) can be generalized by regarding x_{av}, Δx, T_{av} and ΔT as pertaining to a pair of adjacent geophones. In this case ΔT is referred to as the "step out time" or "normal moveout". With a multi-geophone spread many values of the quantities can be measured and a mean taken. In the form of equation (8.42), this approach is often called the "TΔT" method.

Weathered Layer Problems

The top layer in a region is frequently weathered sufficiently to give it a velocity which is very significantly less than that of all deeper layers. Unless allowance is made for this low velocity layer when travel times are converted to depths, the effect would be to make it appear as if the reflecting horizons are deeper than they really are.

If the weathered layer was of uniform thickness there would be no problem since the depth error would be constant for all geophones. However, the thickness of the weathered layer is often highly variable and the effects of an undetected variation in thickness of this layer might be misinterpreted as structure in the reflecting horizon. For instance, if the velocity in the weathered layer is only one tenth the mean velocity in the subsequent layers, a 5 m variation in the thickness would, in terms of reflected travel time, be equivalent to a 50 m variation in depth of the reflecting horizon. A 50 m structure is a highly desirable target, so it would be a pity to drill a 1 or 2 km deep hole only to find that the suspected structure was due to a variation in thickness of a weathered layer which is only a few tens of meters thick.

In all cases, therefore, the individual geophone data must be corrected for elevation differences and variations in weathered layer thicknesses. This is usually done by referring all data to common datum level which is below the weathered layer. Variation in weathered layer thicknesses can sometimes be obtained from first arrival (refraction) data on the reflection records; if not then separate refraction experiments may be needed using close geophone separation because of the detail required.

In most cases the shot point is placed below the weathered layer by drilling a relatively shallow hole. This avoids the need for a correction at the shot point end of the spread, halves the expected energy absortion, which is high in the weathered layers, and gives some control on the thickness of the weathered layer.

Recording Techniques

Wide band FM, or digital tape recordings are now quite commonly used. Visual recorders are used in the field to monitor the results as the program progresses, but the bulk of the filtering and detailed analysis is carried out at a central office. However, some methods of attempting to improve the message to noise ratio, which are really forms of filtering, are still used in the field. These will be discussed later.

With modern recording techniques, the various corrections are often a simple knob twisting black-box affair for analogue recording; for digital data the black-box is replaced by a subroutine. For instance, the elevation and weathering corrections will be constant for any given geophone channel so that a simple subtraction network, adjustable for each channel, can be used. On the other hand, for normal moveout the correction not only varies from channel to channel but varies within a given channel so that a subtraction network which is a function of time is required for each channel. The conversion of the two way travel times to depth requires a knowledge of the velocity depth function before a proper section can be presented; correction for this may require a digitized form of the velocity depth function or, for analogue techniques, a curve follower.

Although all these corrections are simple to make in principle, care has to be exercised in complex areas. For instance, in tectonically disturbed areas, which means there may be a number of discontinuities which dip in azimuth as well as in the direction of the profile, the subsurface reflecting points are generally not even in the vertical plane containing the profile. Instead they are displaced to the side by an amount dpending on the dip, depth and velocity-depth function. These reflection points therefore have to be "migrated" to this time position in space before a proper section can be drawn.

Presentation of Records

There are a number of ways of presenting the data both before and after the corrections have been put in. Many competing claims are made for the advantages of a particular method but, as in so many other cases of data manipulation, none of them gives more information than is in the original data and the chief merit is to give something that is easily interpretable to the eye. Since we are all trained in slightly differing traditions, one form of presentation may be more readily understood by one person than another.

The "wiggle trace", (Figure 8.15(A), was the industry standard form from the early days of seismic exploration. Because signal amplitudes have such a wide dynamic range and various forms of variable amplifications were used so that the traces could be followed at all times, some information was lost. This led to the development of the "clipped trace" method where those signals which exceeded a certain amplitude were clipped, Figure 8.15(B); this is a crude form of what is now called the "variable area" presentation, Figure 8.15(C), where an area which is roughly proportional to the signal amplitude is blacked in. The "clipped trace" alone still resulted in loss of information and the "nested trace" was developed in which the clipped portion of the trace was nested; that is when the signal amplitude exceeded a given amount, the base line was shifted the same amount and it is possible to do this a number of times, Figure 8.15(D). The improved separation of complex wave forms can readily be seen.

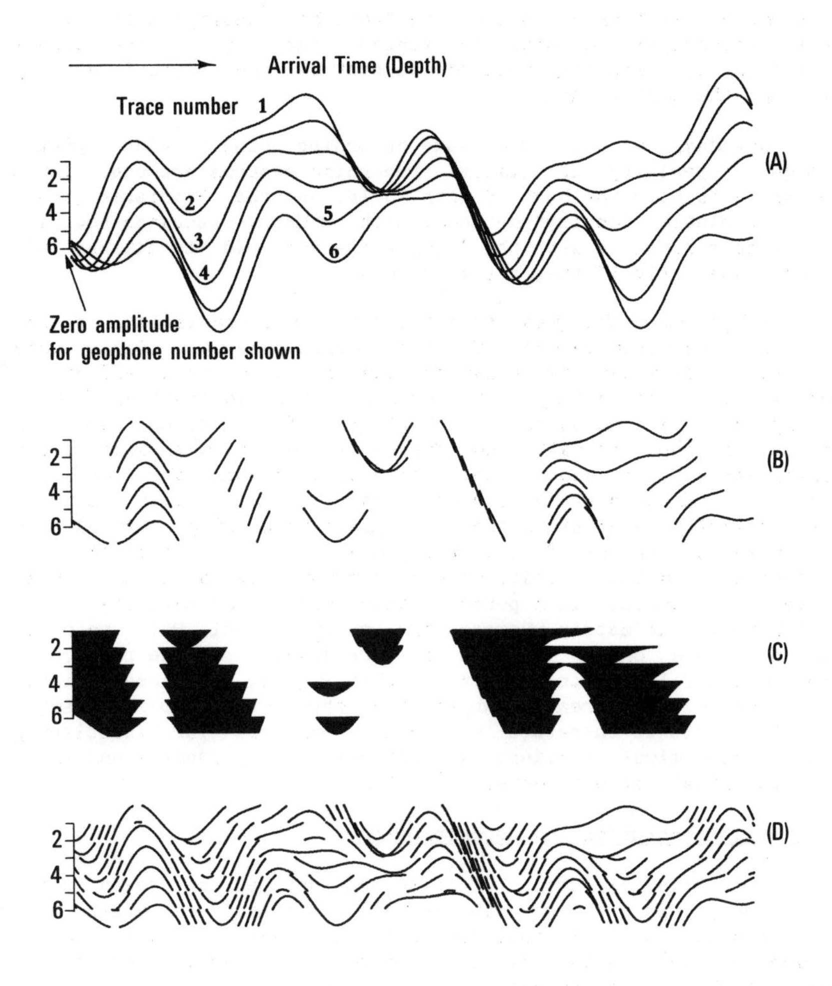

Figure 8.15. Schematic of various methods of presenting records (A) "Wiggle" trace, (B) Clipped trace, (C) Variable area (D) Nested Trace

This method is a crude form of the "variable density" method where the photographic density is varied according to the signal amplitude by using the amplitude to control the current density through the lamp filament.

Nowadays, with the wide band recording using a wide dynamic range, especially if digital recording systems are used, it is possible to have the best of both worlds and the "variable area" and "variable density" methods, both industry standards now, are frequently combined with the "wiggle trace"; old traditions die hard, especially if they are good ones.

All these techniques are related to the presentation of the travel time distance data of the compressional wave. Since large volumes of data can now be manipulated with ease, much effort is being put into attempts to extract more information from an analysis of various other characteristics of the signal such as phase of the arrivals and frequency content, with much of the information being colour coded into bands, in much the same way that contour maps are colour coded to bring out certain features. Of particular interest is the development work being carried out to make better use of shear wave data. In this context the main interest is in the velocity of the horizontally polarized shear wave (SH) which, in a porous medium, depends principally on the velocity in the matrix whereas the compressional wave velocity depends more on the fluid in the pore spaces. If the pores of a medium are filled with gas, rather than liquid, the compressional wave velocity decreases while the shear wave velocity changes little; this is therefore a useful technique for identifying which geological sections are potential gas producers and which are potential oil producers.

Filters, Convolution and Deconvolution

Some brief discussion of filtering must be undertaken since not only is there considerable relevance to modern day electronics, but it is possible to obtain useful information by regarding the earth layers themselves as filters. Furthermore, many of the techniques used in seismology are commonly used in other areas of geophysics to enhance the message to noise ratio.

It hardly needs to be stated that a filter is a device that removes some of the information, usually unwanted, from a signal, Figure 8.16(A).

In electronics a filter can be described in one of two ways:

(a) with amplitude and phase curves by means of which the output from any input can be described. These curves are in the "frequency domain" and represent the "transfer function" or ratio of "output" to "input", Figure 8.16(B).

Distortion occurs if different frequencies have different phase shifts.

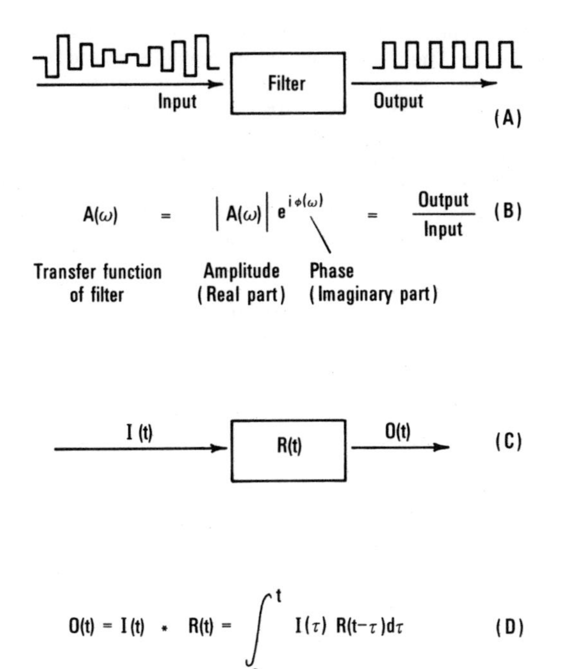

Figure 8.16. Illustrating the function of a filter. (A) and (B) for amplitude and phase approach; note that the long period (low frequency) component has been removed in this example and that a phase difference might exist between the input and output high frequency component. (C) and (D) for impulse response approach; R(t), the impulse response of the filter, is convolved with the input to give the output.

(b) in terms of the "impulse response", which is in the "time domain" of the filter. That is, an impulse is applied to the filter and the output is observed, the results being a time plot similar to that of a seismogram.

Obviously the two methods of describing the filter are not independent and the response curves of (a) can be derived from those of (b) and vice versa, but one method of description may be of more practical use than the other in solving a particular problem. For instance, the amplitude response curve alone is useful if our only interest is in how good a given filter is in discriminating against an undesired portion of the signal in a particular frequency range; that is, we are not interested in the distorting effect the filter has in the frequency range of interest but only in the fact that a signal is there. A practical example would be the use of travel-time distance curves. The important thing here is the time of arrival of the seismic disturbance; it is not very important whether the disturbance appears to arrive as an initial compression or an initial rarefaction, which is 180° out of phase with a compression. However, to determine how faithfully the filter reproduces the wanted signal, the impulse curve, or both the amplitude and phase curves, must be used.

Since a seismic experiment consists of an impulse (at the shot point), and the layered earth can be regarded as a filter, we will concentrate of the impulse curve.

An impulsive input gives a transient output wave-form shape which depends on the filter characteristic whose impulse response can be denoted functionally as R(t). There are many types of filters but we are interested only in linear filters, principally because non-linear filter theory is so complex. A linear filter is one which gives a single output O(t) for a single input I(t),

and can be thought of as a black box with an input and an output, Figure 8.16(C). The impulse response belongs to the filter and it therefore belongs in the black box.

O(t), the output (seismogram) will be a function of I(t), the input (shot), and the impulse response, R(t), of the filter (earth layers). This can be obtained mathematically by "convolving" I(t) with R(t) that is

$$O(t) = I(t) * R(t) = \int_{\tau=0}^{t} I(\tau)\, R(t-\tau)\, d\tau \qquad (8.43)$$

where the integral defines convolution and τ is the two way travel time in the reflection problem. It can therefore be seen that filtering can be regarded as the mathematical process of convolution which has the vital property of being commutative, that is, $I(t)*R(t) = R(t)*I(t)$, and it is not important which of the two wave forms are considered to be the input and which is the impulse response of the filter.

In electronics, it is relatively easy to measure I(t), O(t) and R(t). However, in seismic exploration although O(t) (the wiggle trace record) is measured relatively accurately, it is done only after the input signal has passed through several filters (layers); if information is required about the response characteristics of these filters, some reasonable assumptions about I(t) (the shot impulse waveform) must be made before analysing the output. Fortunately, modern methods of recording and analysing data have made it possible to consider this reverse operation of convolution, known as deconvolution, that is, to reduce the reflection seismogram to a set of impulse responses which represent the various layers.

This means that for a layered earth, a reflection event can be regarded as a multiple interference phenomenon rather than a single event at a discontinuity. Expressions containing the reflection and transmisstion coefficients of each discontinuity can be derived which have the mathematical form of a convolution and straightforward mathematical procedures can then be used for convolution and deconvolution.

Representing the layers by impulse responses is very useful in some stratigraphic studies where changes in character of a particular seismic reflection may be diagnostic of changes in lithology or thickness.

With the continuous velocity log used in boreholes it is possible to obtain an accurate and detailed log of changes of velocity in a given stratigraphic section. It should therefore be possible to calculate the reflection response of that model to a seismic input whose waveform is known approximately. The resultant seismogram is called a theoretical or synthetic seismogram.

There is little point deriving synthetic seismograms in areas where the stratigraphy is already well known, but it can be of value in planning a program to obtain detail on something specific. For example, a prior study of synthetic seismograms may indicate that an unusual frequency range of seismic input will yield more useful information than that obtained using the standard range; this might require modification of recorder speeds and recording times, both cumbersome jobs if done by trial and error in the field.

Energy Injection

Many methods of energy injection are available, the most common one being the explosive charge. Sometimes much experimentation is required in an area before an optimum is found. The ideal pulse should have a strong downward component, with little upward component. This can be obtained by using special patterns of charges and represents a form of filtering the input. Geophone patterns for explosive inputs will be discussed later.

In some areas, it is not possible to use explosive charges and various other forms of energy input must be used. These usually have a much smaller impulse energy than explosive charges and various techniques are used to improve the message to noise ratio.

In the weight drop method, the random noise effects are

reduced by summing the results from several drops. Coherent signals, of which the output response to the geophones might be only one, augment each other while random noise cancels out.

To help remove the effects of coherent noise such as Rayleigh waves (ground roll), the weight is moved in various patterns and large area geophone patches are used. The principle behind this form of filtering is shown in Figure 8.17. The reflection event from depth arrives at all six geophones simultaneously and at precisely the same time for each weight drop as does the Rayleigh wave which travels along the surface. The signal for the three geophones on the maximum of the Rayleigh wave, will always be the sum of those due to the Rayleigh wave and the reflected event.

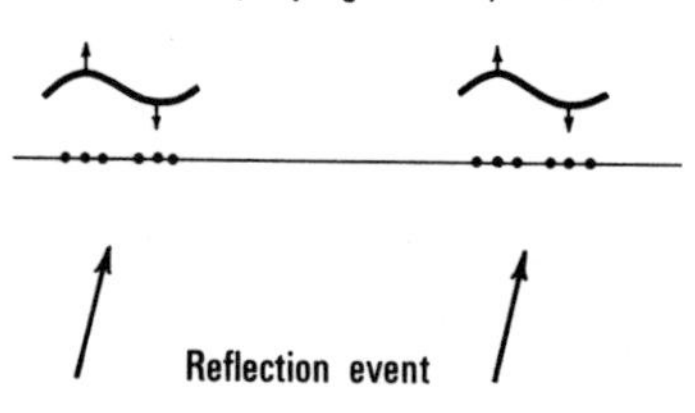

Figure 8.17. Illustrating principle of reducing coherent noise from Rayleigh wave.

This is so for all weight drops so that the Rayleigh wave signal is unwanted coherent noise. The wavelength of the Rayleigh wave can be calculated and other geophones (three in Figure 8.17) are placed so that they are on the minimum of the Rayleigh wave. Therefore, as the first three are moving down (negative signal), the other three are moving up (positive signal) and if the signals from all six are added, the Rayleigh wave signal tends to cancel while that from the reflection event is augmented. Cancellation of the Rayleigh wave noise is never complete because there is a finite time required for it to pass over the geophone patch, and frequency and velocity filtering is used to reduce the noise still further.

Two types of weight drop patterns are shown in Figure 8.18.

In a typical weight drop experiment, the weight may be as heavy as 2000 kg and be dropped from 3 m. On soft ground, coupling plates might be used. Coupling plates are also used if it is required that shear waves be generated; in this case the weight is linked to a pivot so that the "hammer" blow is applied horizontally to the coupling plate.

The mechanical method of injecting energy is used on a much smaller scale in engineering studies of the overburden and near

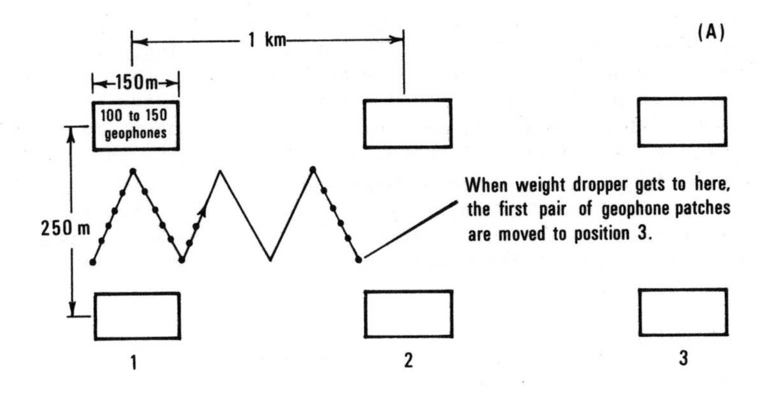

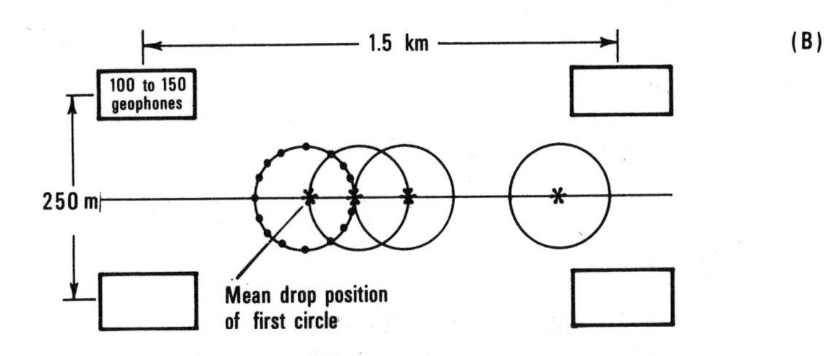

Figure 8.18. Two patterns used in the weight drop technique (A) Saw-tooth or transposed in line (B) Overlapped circles.

surface layers; in such studies a sledgehammer is used to create the seismic disturbance and the detecting and recording instrument can be made quite portable.

The chief advantage of the weight drop technique is that several thousands of ray paths can be summed to average the odd effects that might be obtained from a single ray whose path may traverse a significant inhomogeneity. The chief disadvantages are (a) that refraction data are not available and the weathered layer correction may be unreliable, and (b) high frequencies are virtually absent.

Another form of energy injection consists of an electromagnetic vibrator applied to the earth's surface, either singly or in synchronized multiples. The signal is swept through a predetermined frequency range which depends on the area, and lasts about 6 or 7 seconds.

Because of the long signal time, the response signals are also long and therefore overlap. The shape of the input signal is accurately known and is used to operate on the output signal, which is also accurately known, to separate the individual events. This operation is similar to deconvolution and is known as cross correlation; the process of searching for the signal, to obtain a zero time, is known as autocorrelation. The correlogram, the result of this cross correlation, is recorded on a magnetic tape and is analogous to the more usual field seismogram. Geophone arrays and source patterns are similar to those for the standard weight drop technique.

The chief advantage of the method is that it can be used on soft and difficult ground, such as permafrost areas, where other methods fail.

Geophone Arrays

Continuous profiling. The reflection method is only used for detailed mapping of subsurface structures. A typical continuous profiling procedure is shown in Figure 8.19(A). A shot is fired at S_1 with an equal number of geophone locations on each side. Some of these locations may consist of patches of geophones such as in Figure 8.17 with the geophone separation in a patch being small compared with position separation. Alternatively, to help reduce ground roll noise, the signals from adjacent traces (or up to three or four traces) may be summed.

The shot point is moved to S_2, the left hand set of geophones placed on the right of S_2 and the process is repeated; to speed things up, a third string of geophones may be employed and previously layed out to the right of S_2 so that the first string is layed out starting one spread length away from S_2 after S_1 has been fired.

Multifold techniques are often known as common depth point (CDP) or common reflection point (CRP) techniques.

With many of the previously discussed approaches, there has been an attempt to enhance the message to noise ratio by using multiple impulses, large geophone patches etc. Sometimes the patches become so large and the procedures so complicated and sample so large an area that they tend to obscure the very detail they were originally designed to give.

With common reflection point techniques, the geophone arrays, shot point locations and the movement of these in profiling are designed so that reflection events appearing in certain channels are always from the same reflecting point of a given horizon. This effectively standardizes the reflection characteristics of the reflection point, an awkward variable at the best of times. There is some illogicality in this, as in most seismic work, since

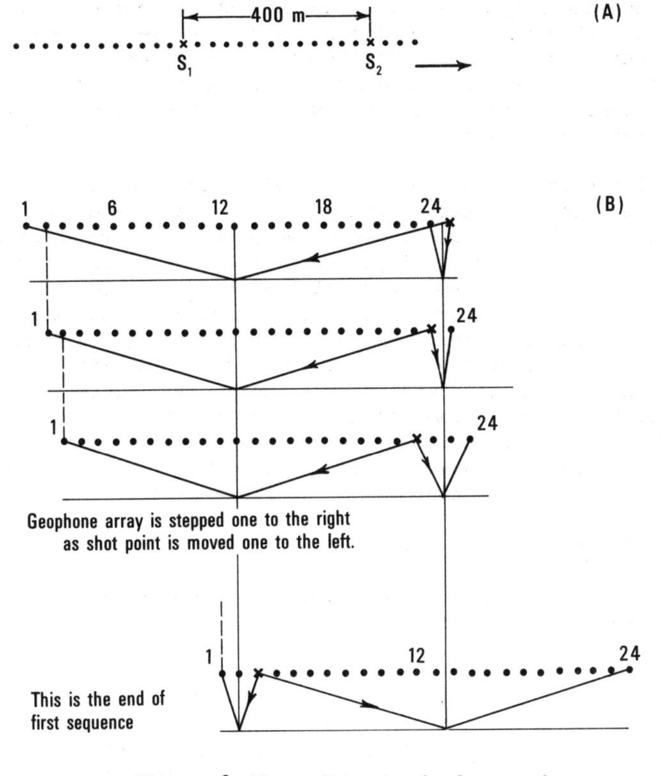

Figure 8.19. Some typical geophone and shot point geometries for reflection methods. (A) Continuous inline profiling (B) Continuous multifold profiling; second sequence starts with shot point on extreme right again.

it assumes the reality of rays, or at least the exactness of ray theory as opposed to the correct wave front theory.

There are various n-fold methods, n being the number of shots giving rise to reflections from the common point. n-fold is sometimes called n x 100% coverage.

Twelve fold (1200 percent) coverage is now quite common and has the big advantage that the reflections, at different angles, are all recorded in the same channel and it is therefore relatively easy to add the information arriving from a given reflection point but due to various shots. A typical procedure for profiling is shown in Figure 8.19(B).

When stacking traces, they are first corrected for normal moveout, elevation variations, etc. Noise effects with a dependance on offset different from that of primary reflections - such as multiple reflections, surface waves, diffractions, refractions, etc. will be attenuated relative to the primary reflection.

All the seismic methods are readily adapted to oceanic or lake use, the principle change being in the energy sources. Geophones are located on streamers held below the water at a depth of a few meters. Impulses are periodically applied by capacitative discharges, gas explosions or even conventional charges, and recordings made repetitively as the vessel moves. Multifold techniques are especially adaptable for oceanographic work since with a suitable choice of geophone spacing in the towed string the repetitive signal source can be synchronised with the speed of the ship. These techniques are particularly useful since they reduce the problem of reverberation - an effect due to multiply reflected waves; in this case, the velocity of the vessel has to be dependent on the geophone separation.

SUGGESTIONS FOR FURTHER READING

Dobrin, M.B., 1976. Introduction to geophysical prospecting. McGraw-Hill, New York.

Grant, F.S. and West, G.F., 1965. Interpretation theory in applied geophysics. McGraw-Hill, New York.

Hobson, G.D., 1970. Seismic methods in mining and ground water exploration, pp. 148-176 in Mining and Groundwater Geophysics, ed. L.W. Morley. Ec. Geol. Rep. No. 26, Geol. Surv. Canada, Ottawa.

Northwood, E.J., 1967. Notes on errors in refraction interpretation, pp. 459-465 in Seismic Refraction Prospecting, ed. E.W. Musgrave, Soc. Expl. Geophys. Tulsa, Okla.

Taner, M.T. and Sheriff, R.E., 1977. Application of amplitude, frequency, and other attributes to stratigraphic and hydrocarbon determination, pp. 301-327, in Seismic Stratigraphy - applications to hydrocarbon exploration, ed. C.E. Payton, Am. Assoc. Petrol. Geol. Tulsa, Okla.

9

Radioactive methods

BACKGROUND ON ATOMS AND ISOTOPES

The structure of an atom, particularly the nucleus, is so complex that even now neither the structure nor the short range forces which hold the nucleus together are completely understood. However, there are a few things which can be said about atoms, amongst which are:

(a) the diameter is about 10^{-8} cm
(b) the diameter of the nucleus is about 10^{-12} cm
(c) the nucleus consists of nucleons, i.e. neutrons plus protons
(d) the mass of the neutron is only slightly greater than the mass of the proton
(e) the mass of the proton is eighteen hundred times the mass of the electron
(f) protons have an electrical charge of +1 while electrons have a charge of -1
(g) electrically neutral atoms have the same number of electrons, in various orbits, as there are protons in the nucleus.

The largest naturally occurring nucleus is that of Uranium-238 which contains 92 protons; the number 238 refers to the atomic mass using the proton (or neutron) as the unit of mass. Since there are only 92 protons in the nucleus and its mass is 238, this means there must be 146 neutrons in the nucleus as well. The number of protons or positive charges in the nucleus determines the Atomic Number Z. Thus Uranium-238 is usually shown as ${}^{238}_{92}U$. An atom is completely described by its atomic mass and atomic number and use of a letter indicating the colloquial name of the element is really irrelevant.

Neutrons can be written as ${}^{1}_{0}n$ and electrons as ${}^{0}_{-1}e$ although n and e alone are frequently used since there is some inconsistency in the use of subscripts and superscripts compared with their use for atoms.

In the case of many naturally occurring elements, particularly gases, it was known for a long time that the atomic weight was not always an integral number. J.J. Thomson found that the elements consisted of a mixture of atoms whose nucleii contained the same number of protons but different numbers of neutrons. That is they had the same atomic number but different atomic masses. Since it is the atomic number Z (or distribution of electrons around the nucleus) which determines chemical properties, these atoms are described as isotopes of the same element. Because the chemical properties are identical, isotopes cannot easily be separated by chemical methods, only by physical methods.

Spontaneous Disintegration

Many naturally occurring elements are found to emit radiation spontaneously; the emission is unaffected by any chemical or physical operation normally performed on an element and must therefore be a property of the nucleus.

Three readily detected types of disintegration products are emitted.

(a) α-particles which are helium nuclei, carry a +2 positive charge ($^{4}_{2}He^{++}$). They produce very intense ionization in gases and travel with velocities ranging from 1.4 to 2.2 x 10^{10}cm/sec.

(b) β-particles which are electrons emitted with velocities ranging from 10^{10} cm/sec to nearly the velocity of light (3 x 10^{10} cm/sec).

(c) γ-rays which are really penetrating X-rays of very short wavelength (about 10^{-9} cm). They may therefore be regarded in precisely the same way as light - either as bursts of electromagnetic radiation of very high frequency, or as discrete or quantized energy packets of zero mass and called photons. γ-rays are usually generated in association with the emission of α- and β-particles which leave the nucleus in an excited state which then emits a γ-ray (or rays) to attain a stable state.

In addition to the above, positrons can be emitted, as can neutrons when the much rarer event of spontaneous fission occurs.

Pattern of Decay

Consider a parent atom $^{m}_{n}A$ which now emits an α-particle. The mass has been reduced by 4 units and the charge by 2 units. This can be described by the following equation

$$^{m}_{n}A \rightarrow \, ^{m-4}_{n-2}B + \alpha \qquad \text{or} \qquad ^{m}_{n}A \rightarrow \, ^{m-4}_{n-2}B + \, ^{4}_{2}He \qquad (9.1)$$

Since the atomic number has changed, the remains of the atom must be a new element. If this new element itself decays by the emission of a β-particle, the change can be represented by (9.2).

$$ {}^{m-4}_{n-2}B \rightarrow {}^{m-4}_{n-1}C + \beta \quad \text{or} \quad {}^{m-4}_{n-2}B \rightarrow {}^{m-4}_{n-1}C + {}^{0}_{-1}e \tag{9.2} $$

If element C now disintegrates by losing a β-particle, we have

$$ {}^{m-4}_{n-1}C \rightarrow {}^{m-4}_{n}A + \beta \quad \text{or} \quad {}^{m-4}_{n-1}C \rightarrow {}^{m-4}_{n}A + {}^{0}_{-1}e \tag{9.3} $$

Although the remaining atom in (9.3) has a different mass from the parent atom ${}^{m}_{n}A$, it has the same atomic number and so must be an isotope of element A. The complete set of disintegrations can be represented by (9.4)

$$ {}^{m}_{n}A \rightarrow {}^{m-4}_{n}A + \alpha + 2\beta \quad \text{or} \quad {}^{m}_{n}A \rightarrow {}^{m-4}_{n}A + {}^{4}_{2}He^{++} + 2\,{}^{0}_{-1}e \tag{9.4} $$

The essential point in writing any equation for radioactivity is that the sum of the mass numbers and the sum of the atomic numbers must be the same on both sides.

Decay Laws

Experimentally, it can readily be determined that a quantity of radioactive atoms decays at a rate proportional to the number of decaying atoms present (9.5)

$$ \frac{dA}{dt} = -\lambda A \tag{9.5} $$

where A is the number of atoms of radioactive isotope ${}^{m}_{n}A$ present now and λ is the constant of proportionality called the decay constant; the negative sign simply indicates that the number of atoms of ${}^{m}_{n}A$ decreases as time increases.

Separating the variables and integrating both sides gives (9.6)

$$ A = A_o e^{-\lambda t} \tag{9.6} $$

where A_o is the number of atoms of ${}^{m}_{n}A$ originally present at time t = 0, and t is the time between then and now.

Physically, the decay constant can be regarded as a probability; if its value is, say, 10^{10} sec^{-1} this means that if a particular atom is observed for one second, there is only one chance in 10^{10} that it will spontaneously disintegrate - or, more accurately since the distribution is Poissonian, if one thousand atoms were watched for one second there is only one chance in 10^7

that one of the atoms will disintegrate.

An alternative concept is that of half life. Let T be the time required for half the original number of atoms, A_o, to disintegrate, then from (9.6)

$$\frac{A}{A_o} = \frac{1}{2} = e^{-\lambda T} \tag{9.7}$$

so that

$$T = \frac{\ln 2}{\lambda} = \frac{0.693}{\lambda} \tag{9.8}$$

T is called the half life of the atom and, again in terms of probability, it can be interpreted as meaning that if, for instance, $T = 10^9$ years, then if a particular atom of that element is observed for 10^9 years, there is a 1 in 2 chance that it will disintegrate while being observed.

DECAY SERIES OF INTEREST

There are many elements in nature which undergo spontaneous disintegration and many have been used to give information on the absolute ages of rocks and minerals. Only three elements are of significant interest in exploration work, Uranium, Thorium and Potassium. Two isotopes of Uranium (235 and 238) should be considered although for the sake of simplicity only ^{238}U, which is 137 times more abundant than ^{235}U, will be referred to. Simplified schemes for ^{238}U, ^{232}Th and ^{40}K are shown in Figure 9.1. The actual decays are more complicated than shown since many of the intermediate daughter products have more than one form of decay; in this case the total probability that the daughter product will decay is the sum of the probabilities of the alternative decays occurring. For simplicity, only those branching decays greater than 10% (0.1) of the total are shown.

Secular Equilibrium

In a series with a chain of decays, such as for ^{238}U, the simple decay equation, (9.5), is complicated by the fact that as an intermediate daughter, I_n, decays it is also being continuously created by the decay of the immediately preceding daughter, I_{n-1}. The form of the differential equation then becomes

$$\frac{dI_n}{dt} = \lambda_{n-1} I_{n-1} - \lambda_n I_n \tag{9.9}$$

where λ_{n-1} and λ_n are the decay constants of the (n-1)th and nth daughters present in amounts I_{n-1} and I_n at the time of measurement.

The solution to (9.9) is

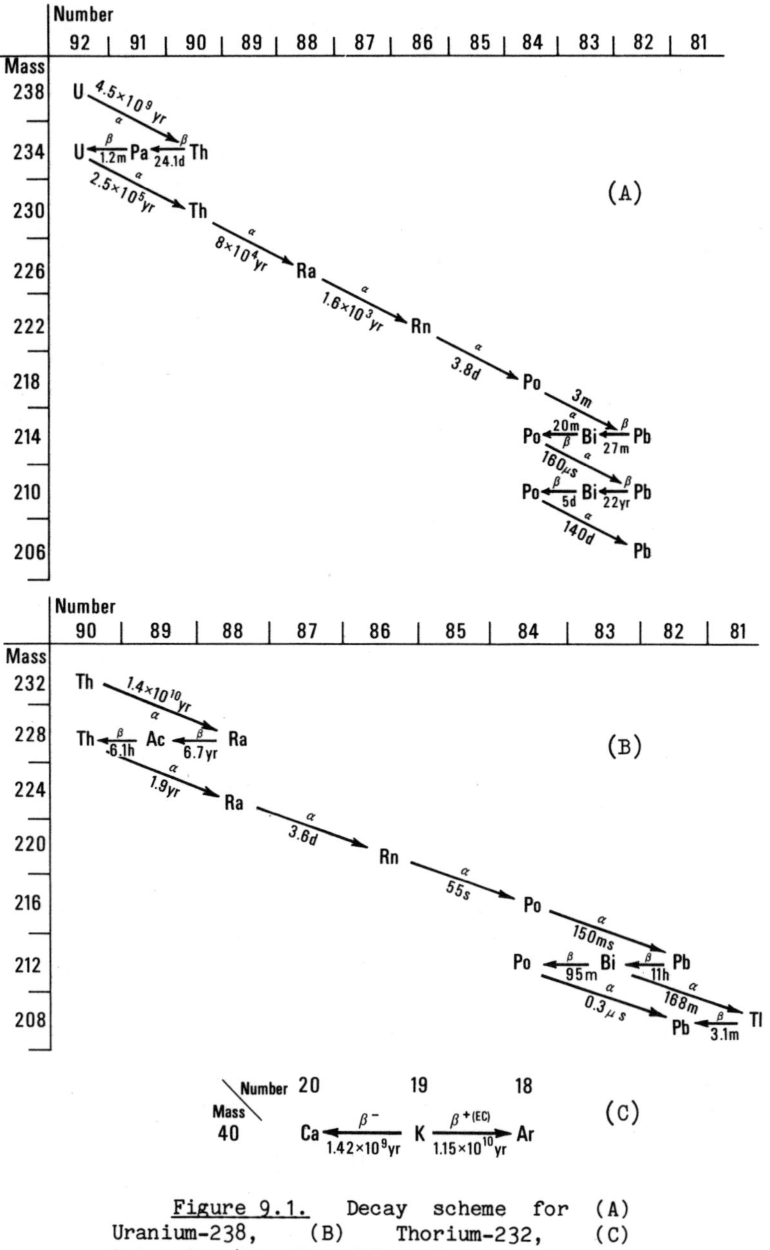

Figure 9.1. Decay scheme for (A) Uranium-238, (B) Thorium-232, (C) Potassium-40. Branching decays of less than 10% are not shown here.

$$I_n = \lambda_o\lambda_1 \cdots\cdots \lambda_{n-1} P \left\{ \frac{1}{(\lambda_n-\lambda_o)(\lambda_{n-1}-\lambda_o) \cdots (\lambda_1-\lambda_o)} + \cdots \right.$$

$$\left. \cdots + \frac{e^{(\lambda_o - \lambda_n)t}}{(\lambda_o-\lambda_n)(\lambda_1-\lambda_n) \cdots (\lambda_{n-1}-\lambda_n)} \right\} \qquad (9.10)$$

If $\lambda_n > \lambda_o$ then for $\lambda_n t >> 1$, which means that $0.693\ t/T_n >> 1$ and therefore that several half lives have elapsed (from equation 9.8), we obtain

$$I_1 = \frac{\lambda_o P}{\lambda_1-\lambda_o} \; ; \quad I_2 = \frac{\lambda_o\lambda_1 P}{(\lambda_2-\lambda_o)(\lambda_1-\lambda_o)} \qquad (9.11)$$

If in addition $\lambda_n >> \lambda_o$, which means that the half life of the parent is very much larger than the half life of any of its daughters.

$$I_1 = \lambda_o P/\lambda_1 \; ; \quad I_2 = \lambda_o P/\lambda_2 \quad \text{or} \quad I_n = \lambda_o P/\lambda_n$$

from which we get

$$\lambda_1 I_1 = \lambda_2 I_2 = \cdots = \lambda_n I_n = \lambda_o P$$

or

$$I_1/I_n = \lambda_n/\lambda_1 = T_1/T_n \qquad (9.12)$$

When this simplification holds, the series is said to be in radioactive, or secular, equilibrium and means that, on average, for each decaying parent atom, one of each of the daughter atoms also decays; the number of atoms of each intermediate daughter atom will be directly proportional to its half life or inversely proportional to its decay constant. In practice, for such a statement to be true the system must be closed, that is neither parent nor daughter can be lost or gained from the system, and the half life of the parent should be considerably longer than any of its daughter products. Fortunately the latter statement is true for both Uranium series and the Thorium series but whether or not a source or sink exists depends on geological conditions.

From Figure 9.1(a) it will be noted that Radon-222 occurs in the series; this isotope has a half life of a few days and is a gas at normal temperatures and pressures so that it is possible that loss of Radon occurs, particularly in fractured rocks. This can lead to several problems in age determination work and exploration work, but it can have advantages as well.

γ-RAY SPECTROMETRY

γ-rays interact with matter in a number of ways the principal ones of concern to us being the photo-electric effect, the Compton effect and pair production.

Photo Electric Effect

Each γ-ray (photon) has an energy E which depends only on its frequency (ν) or wavelength (L) and is given by

$$E = h\nu = h(c/L) \tag{9.13}$$

where h is Planck's constant and c is the velocity of light.

A single photon can interact only with a single electron in the atom it hits.

Since photons have no rest mass and travel with the speed of light their energy is entirely kinetic. If on collision with an electron the photon is completely stopped, it must cease to exist and, since energy must be conserved, all the photon energy is transferred to the electron it hits. If the energy exceeds the binding energy of the electron, the electron is ejected with a kinetic energy (K.E.) equal to the excess of the photon energy over the binding energy. Therefore, for any given γ-ray energy (or, from 9.13, frequency), there is a well defined K.E. for the photo-electron. If the K.E. of the electron is measured, and the original binding energy is known, the energy of the incident γ-ray can be found.

Fortunately, the converse effect holds. When an electron loses K.E. it can create a photon; if it can be arranged that this photon has an energy which corresponds to an E.M. frequency in the visible range, a flash of light would be observed (assuming the photon density is great enough for the eye to detect). In practice there are usually not sufficient photons for visual detection and an amplifier with associated electronics can be designed which essentially counts the number of photons. This is the principle of the photomultiplier and Thallium activated Sodium Iodide (NaI[Tℓ]) crystal detector much used in exploration work.

Compton Effect

A photon does not necessarily give all of its energy to an electron; if only part of the energy is given to the electron it leads to the scattering of E.M. waves by the charged particles of the target material. This type of quantum scattering is known as the Compton Effect or Compton Scattering.

In the photoelectric effect, it was assumed that the incident photon was completely stopped and therefore anihalated, so that only conservation of energy was considered. For the Compton effect the impact is examined in the light of an elastic collision so that the conservation of momentum must be considered, Figure 9.2.

If hc/L is the energy and h/L is the momentum (from equation

9.13) of the photon striking a stationary electron, e_0, and sufficient energy is transferred to exceed the binding energy, the electron is set in motion as a photo-electron (or recoil electron).

From the equations of conservation of energy and momentum we can predict

(1) that the diverted photon must have a different energy from the incident photon and therefore must have a different frequency, equation 9.13.

(2) the angles of scattering of both the scattered photon and the photo-electron; these latter predictions are of little interest to the exploration geophysicist.

There are complications in this simple picture since the electrons are, in general, neither at rest nor free, but it is usually the electrons in the outer shells, that is those with the lowest binding energies, that carry the brunt of the bombardment.

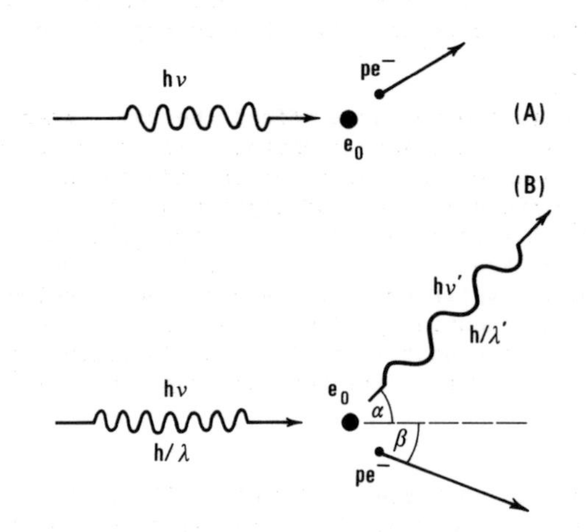

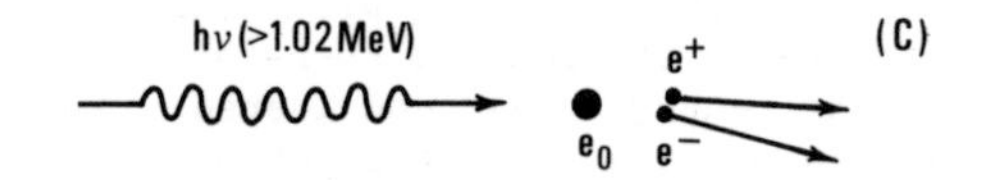

Figure 9.2. Illustrating the basics of (A) photoelectric effect, (B) Compton scattering, (C) pair production.

The important point is that frequency and direction of motion of the scattered photon depends on many things; in one sense the process can be regarded as the instantaneous destruction of one photon of a given energy, and the creation of a new photon of lower energy, the energy difference being imparted to the recoil electron.

Pair Production

The only other sort of reaction that need be considered, and then only briefly, is pair production. The previous reaction essentially dealt with the conversion of electromagnetic energy into the kinetic and potential energy of particles and vice versa. However, all the energy can be converted into pure matter provided the energy of the incident photon exceeds a certain limit.

If this happens, it can be seen intuitively that another form of conservation must be considered - that of charge. Since the electron has the least rest mass of any elementary particle it should be the easiest to create from energy; but to conserve the charge a positron most be created simultaneously.

The minimum energy required to create an e^+-e^- pair is simply $2\,m_0c^2$ where m_0 is the rest mass of the electron and c is the velocity of light. Substituting appropriate values gives

$$2m_0c^2 = 1.02\ \text{MeV} \tag{9.14}$$

A photon of this energy would have a wavelength of 0.012 Å.

The electron and the positron, which does not last long, have equal energies and will undergo other collisions creating more photons; however, the process is not simple. For example, if a minimum energy positron collides with an electron to create pure energy again, the result is not, as one might first expect, a photon of 1.02 MeV energy; to conserve momentum and energy, two photons (γ-rays) must be created and the minimum energy of these γ-rays is 0.51 MeV. It is possible to conserve momentum by creating more than two photons moving in different directions but the probability of this occurring is rare.

Even though the above three reactions are fairly easy to understand for single interactions, they are highly complex for the typical geological absorption problems.

Fortunately, as may be seen from Figure 9.3, it is only the Compton effect that has considerable significance for the typical geological problem.

γ-RAY SPECTRA

Each radioactive element has a typical α, β and γ-ray spectrum which under ideal conditions could be detected with high resolution. Unfortunately, all particles must pass through some matter before escaping for detection; if the particles are not stopped completely before reaching a detector, there is still so much Compton scatter causing a line broadening effect on the spectrum that a great deal of overlap occurs between lines. Because γ-rays have the greatest penetrating power of any of the decay products they are the only radioactive products useful for field work; α-spectrometry has been used in laboratory studies only.

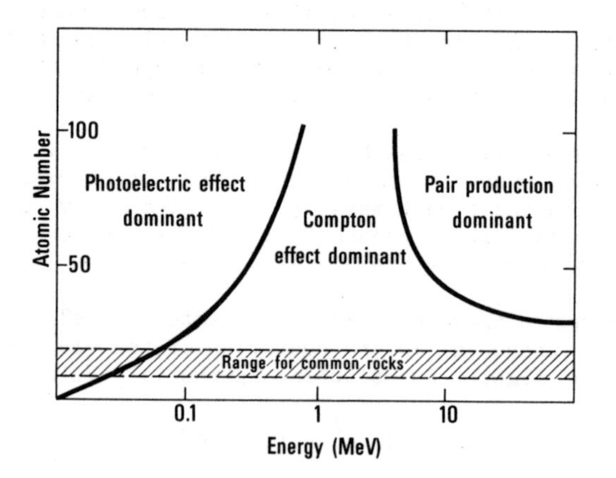

Figure 9.3. Illustrating how γ-ray reactions depend on the atomic number Z and γ-ray (photon) energy. Shaded zone indicates range of Z for common rocks. Since γ-ray energy from the U, Th, and K decays fall in the range from 1 to 3 MeV, the dominant effect in rocks in Compton scatter.

The type of line broadening to be expected can be seen from Figure 9.4 (A) - (C). For the individual element, the height of each line represents the intensity (counts/sec, say) whereas its distance from the y-axis is a measure of the energy.

It can readily be seen that in the ^{232}Th series, there is a characteristic γ-ray of 2.62 MeV, given off by the decay of ^{208}Tℓ fairly late in the decay scheme, which undergoes no interference from decay products of Uranium or Potassium, although it does have a "Compton Tail".

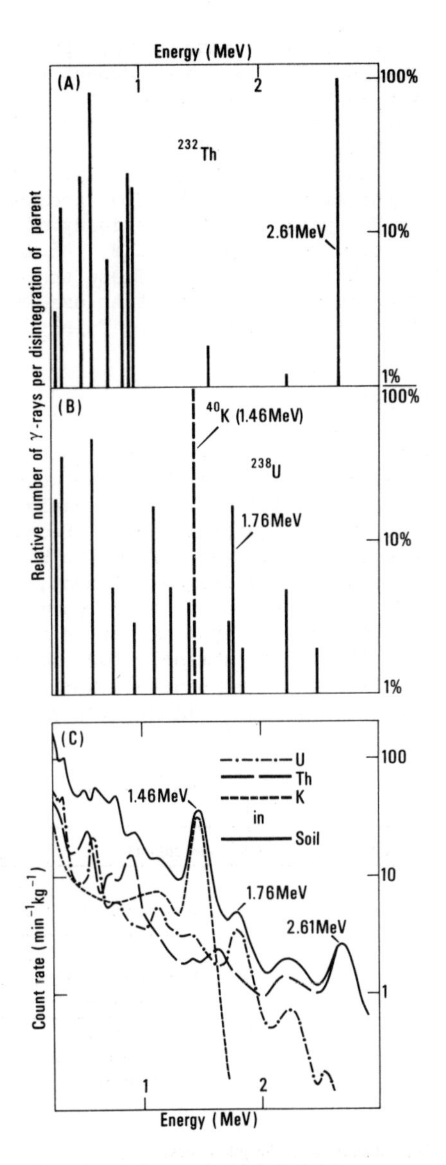

Figure 9.4. Showing how initial line spectra become blurred due to Compton effect. (A) Spectrum for Th; (B) Spectrum for U and single line (dashed) spectrum for K. Only those decays with greater than 2% relative efficiency are shown. (C) Typical spectrum for a soil obtained using a NaI [Tℓ] crystal, showing count rates in Potassium, Uranium and Thorium channels as well as total count rate (solid line).

There is also a fairly intense γ-ray product of 2.19 MeV; but when Uranium and Thorium are present together, as happens almost invariably even though the ratios can vary widely, this energy peak cannot be used since it is almost coincident with the highest peak, 2.20 MeV, of the ^{238}U series. For the same reason, and because it is much less intense, the 2.20 MeV peak cannot be used to determine the amount of Uranium present. To estimate Uranium, the 1.76 MeV peak, due to the decay of ^{214}Bi in the ^{238}U decay scheme, is used. For Potassium the 1.46 MeV γ-ray of the ^{40}K to ^{40}A decay branch is used. These three widely separated energies are chosen since they allow the use of fairly simple pulse height analysis equipment with fairly wide channels (see Figure 9.5).

PULSE HEIGHT ANALYSERS

A pulse height analyser simply sorts out the energies of incident gamma rays; the resolution between adjacent γ-ray energies can be as fine as desired but clearly the higher the resolution the greater the equipment cost. Because of the Compton effect, some of the scattered γ-rays from the higher energies, of either the same element or a different element, will have energies in the channel of a different primary gamma ray; there is therefore little point having anything better than a three or four channel equipment (the fourth channel for total count of all energies) in the field, and some method must be devised to allow for distribution of counts from higher energy to lower energy channels.

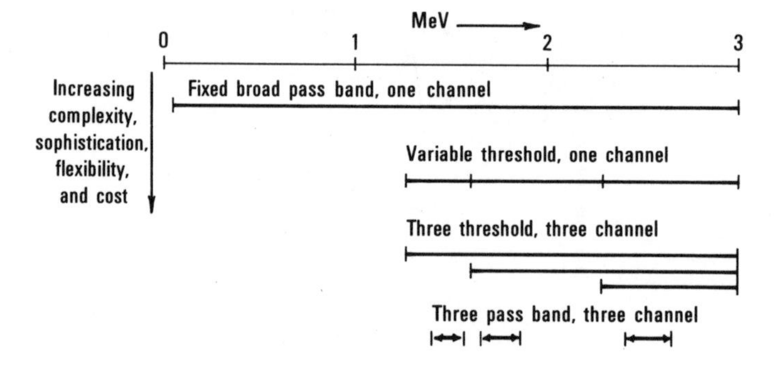

Figure 9.5. Illustrating the various types of γ-ray counting equipment available. The Geiger counter much used by individual prospectors falls in the fixed broad pass band type; typical portable equipment used by exploration companies would have three or four channels (one channel for total counts over a wide range).

Calibration

The instrument is first calibrated with salts of each parent element of known concentrations; several different concentrations of each salt are desirable to check the linearity of the system. Calibration constants are derived and the calibration curve will then give the abundance of each element without interference from the other.

In a practical case suppose that all three of U, Th and K are present. It is usual to start with ^{232}Th because its 2.62 MeV gamma ray, although undergoing line broadening because of Compton scatter, is pure in the sense that none of the counts in that channel can be due to scattered γ-rays from the other elements since all their primary γ-rays are of lower energy. The concentration of ^{232}Th is determined as shown in equation 9.15

$$Th_{ppm} = C_1(Th_{cpm} - BG_{Th}) \qquad (9.15)$$

where Th_{ppm} = abundance of Thorium in ppm

Th_{cpm} = observed count rate in the 2.62 MeV channel

BG_{Th} = background count rate in the 2.62 MeV channel

C_1 = constant related to the system efficiency factor in the 2.62 MeV channel

Since the amount of ^{232}Th present is now known allowance can be made for the contribution of its scattered gamma rays to the 1.76 MeV window count for Uranium, using the calibration constants and correction factors derived from the original calibration (9.16).

$$U_{ppm} = C_2\,[U_{cpm} - BG_u - S_1(Th_{cpm} - BG_{Th})] \qquad (9.16)$$

where U_{ppm}, U_{cpm} and BG_U are the obvious equivalents of the Thorium subscripts, and S_1 is a constant relating to the interference of high energy Th generated Compton scattered γ-rays in the U channel. S_1 is often called the stripping factor.

Similarily the contributions of both Th and U γ-rays scattered into the 1.46 MeV channel for Potassium can be allowed for

$$K\% = C_3\left\{K_{cpm}-BG_k-S_2[U_{cpm}-BG_u-S_1(Th_{cpm}-BG_{Th})]-S_3(Th_{cpm}-BG_{Th}\right\} \qquad (9.17)$$

where S_2 = stripping factor for U generated γ-rays in the 1.46 MeV channel and S_3 = stripping factor for Th generated γ-rays in the 1.46 MeV channel

This process is known as `stripping` since in effect the Thorium spectrum is being strippped from the whole to derive the Uranium spectrum, and both the Uranium and Thorium spectra are stripped from the whole to obtain the Potassium spectrum.

FIELD TECHNIQUES

Disequilibrium and eU

Before describing field procedures, there is one aspect worth discussion. Uranium is estimated from the intensity of the 1.76 MeV γ-ray from the decay of ^{214}Bi. This element occurs late in the series after ^{222}Rn, with a half life of about 3.8 days, and other elements higher in the series of much longer half lives (order of thousands of years). Since Radon is a gas this means that it can be lost from the system, in which case the amount of Uranium will be underestimated. On the other hand if some of the earlier long lived elements are lost from the system, e.g. by leaching, the amount of Uranium will be overestimated; since Uranium behaves as a hexavalent ion in a sedimentary environment, forming a readily soluble uranyl ion, this possibility is by no means unlikely. In such conditions the system is said to be in disequilibrium and to alert a reader as to whether or not the data have been checked by a chemical method the use of eU indicates an indirect estimate of Uranium, usually from a daughter after Radium. Similarly eTh is used to indicate an indirect estimate of Thorium; however, in this case, because the daughter products of ^{232}Th are all so short lived (none exceeding a decade), the Thorium series is nearly always in secular equilibrium.

So that reliable estimates of Uranium are available, it is necessary to check the Uranium value directly by chemical methods, fission track methods or α- or γ-spectrometric methods using high resolution solid state detectors. In the latter approach it is fairly simple to check both the amount of Uranium present and whether the series is in equilibrium in the sample by measuring the relative intensities of the 63 keV γ-ray peak from ^{234}Th (the first daughter in the ^{238}U series and having a half life of only 24 days) and the 47 keV γ-ray peak from ^{210}Pb (which is one of the last three daughters of the ^{238}U series).

Ground Surveys

The equipment usually consists of a 1 to 4 channel spectrometer with a 5 cm to 8 cm NaI[Tℓ] crystal. The crystal is set on the ground and sufficient time allowed for a reasonable number of counts. In principle, for equal standard errors for each measurement, the counts, rather than the count time, should be the same.

Setting the detector on the ground gives what is known as 2π geometry; that is, the detector receives all gamma rays from the

2π sterad solid angle below it although 90% of the count comes from about 60% of this angle.

The data are plotted and contoured, and interpretation made qualitatively. Quantitative treatment is rarely possible for the following reasons.

Although γ-rays have far more penetrating power than α or β rays, even the most energetic γ-ray is stopped by about a 1 m thickness of rock. This means that the detector "sees" little gamma activity from below a depth of 1m and the method might appear to be useless as an exploration tool. Although this may well be true for detailed work, it can still be a useful reconaissance tool since studies have fortunately shown that in areas covered by overburden the soil often correlates well with, and is therefore reasonably representative of, the subsurface rock from which it is usually derived; for barren hard rock areas one makes the simple assumption that the top metre or so of material is representative of that at depth. One of the most promising uses of the γ-ray method is probably in the investigation of soils rather than a search for mineral deposits.

Radon Sniffing

A completely different approach to mapping Uranium concentrations involves the contouring of radon concentration anomalies. ^{222}Rn, having a half life of only 3.8 days does not travel very far before decaying to Polonium-218 therefore sampling the air, or preferably groundwaters, and subsequent geochemical analysis helps to outline an area on the ground which may be worth drilling for the source. However, there are some problems.

As a gas, the path of the Radon from its source to the sampling point may not be consistent from one year to the next, or even from season to season in a given year. For example, in a region subject to winter ice conditions, the Radon may accumulate, and spread, beneath the ice cover during winter and be released to the air and melt runoff during the spring. This gives rise to erratic point to point measurements although with enough measurements some filtering techniques might prove useful. Again, once released into the atmosphere the Radon gas may get carried by winds some distance before being deposited again by rain.

For these reasons, some attention is now being given to using ^{210}Pb as a suitable daughter indicator. This radioisotope has a half life of 22 yr whereas from ^{218}Po to ^{210}Pb the longest half life is 27 m (see Figure 9.1A), with ^{214}Bi, which produces the 1.76 MeV γ-ray usually measured, having a half life of 20 m. A measurement of ^{210}Pb, whether by chemical means or γ-ray spectroscopy, should therefore represent a much longer averaging period. The net result should be to outline a larger but more stable target area.

Helium "sniffing" is a technique similar to "radon sniffing", useful if carefully applied to groundwater.

AIRBORNE METHODS

As usual, the equipment is far more sophisticated, and expensive, than the ground equipment, multichannel pulse height analysers being the rule rather than the exception.

Again, because of the limited range of γ-rays (a few hundred meters in air) the method is used only as a regional geological mapping tool to separate major rock units (if necessary, a finer separation can be attempted by ground methods). Flight height is limited to 200 to 500 m above ground level since at greater elevations, secondary gamma rays from cosmic ray interactions in the atmosphere begin to dominate. The results are contoured using abundance ratios, as well as pure abundances, to try to identify regions of interest for ground follow up.

In general, the Th/U ratio = 4 and the K/U = 10^4 over wide ranges of abundances, but in spite of the geochemical similarities of U and Th most economic deposits of Uranium leave U concentrated relative to Th; the U/Th ratio is therefore usually a better indicator than the U/K ratio.

Using ratios has another major advantage. The total count rate depends on the solid angle "seen" by the detector (as well as on abundances, etc.). In a plane flying at 100 m the ground clearance might vary considerably due to topographic relief. An anomaly in, say, a total count instrument might therefore be due to simple topographic effects. If abundance ratios are used the same angle is "seen" by all channels and the topographic errors therefore tend to be self cancelling for ratios.

One major problem is relating count rates to ground abundances for, in addition to normal stripping, allowance has to be made for the diurnal variation of ^{214}Bi (because ^{214}Bi is derived from ^{222}Rn gas) in the atmosphere with time of day. At about 100 m elevation the abundance is usually a maximum in the early morning and decreases up to 15% as atmospheric mixing occurs. Then overnight, the ^{214}Bi again builds up because of that portion of the Uranium decay scheme from ^{222}Rn gas to ^{214}Bi which involves short lived isotopes (see Figure 9.1(A)). Occasionally, very high ^{214}Bi concentrations occur near the ground following a heavy thunderstorm.

SUGGESTIONS FOR FURTHER READING

Adams, J.A.S. and Gasparini, P., 1970. Gamma-ray spectrometry of rocks, Elsevier, New York.

Bailey, R.V. and Childers, M.O., 1977. Applied mineral exploration with special reference to Uranium, Westview Press, Boulder, Colorado.

Dyck, W., 1972. Radon methods of prospecting in Canada, pp. 212-243 in Uranium prospecting handbook, eds. K. Bowie, L. Davis and I. Ostle, Inst. Min. Met. London.

Dyck, W., Jonasson, I.R. and Liard, R.F., 1976. Uranium prospecting with Rn-222 in frozen ground, J. Geochem. Expl., 5, 115-127.

Szoghy, I.M. and Kish, L., 1978. Determination of radioactive disequilibrium in Uranium bearing rocks, Can. J. Earth Sci. 15, 33-44.

Tanner, A.B., 1964. Radon migration in the ground: a review. pp. 1069-1081 in The natural radiation environment, eds. J.A.S. Adams and W.M. Lowder, U. Chicago Press.

10

Well logging techniques

INTRODUCTION

Nearly all the methods used in surface exploration can be adapted to give a detailed profile of the physical property variations along a borehole.

Well logging techniques are used most commonly in the oil and gas industry, principally because holes are often of large radius and go to great depths; to obtain complete core from such holes is very expensive and collecting chips as the hole progresses is not necessarily definitive. For example, if the hole is progressing at a great rate (in a number of areas a kilometer per week is not uncommon) massive contamination, a mixing of chips from various levels in the hole, can occur. Therefore, we rely greatly on accurate and reproducible results from well logging methods.

Another reason why well logging techniques are little used in the mining industry but much used in the oil and gas industry is the fact that the variation of physical properties of porous sedimentary rocks is considerably greater than in hard rock. The physical properties of interest are often very dependent on the fluids in the pores so that the variations can be used to discriminate between the types, as well as the quantities, of fluids in the pores.

Many techniques are useful only as aids to stratigraphic correlation - the variations in the log of a given property are not analysed quantitatively but the character of the logs from different holes can be used to determine the continuity of lithology.

Frequently the detailed lithology of one borehole is found by complete coring, the character of the log is found and correlated with the lithology, and the the lithology of uncored holes can be determined by a comparison of the logs. The completely cored hole is frequently called a stratigraphic control hole.

However, there are some logging techniques that are quantitative. These are usually used to determine properties such as density, fluid content of pores, hydrocarbon content etc. In this chapter we will deal only briefly with the many types which can be classified broadly as

(a) mechanical
(b) electrical
(c) nuclear
(d) sonic

MECHANICAL TYPE LOGS

Caliper Log. This is a tool with which the diameter of a hole is measured. There are a number of feelers attached to the tool, usually four although to study the corrosion of casing there might be several fingers. The fingers are electro-mechanical devices, held by springs against the wall of the hole, which signal information to the surface. The information from the log is used mainly to estimate the volume of cement that might be required to seal around a collapsed region.

Casing-Collar Locator. This is a scratcher device used to locate precisely where two lengths of collar are joined together.

ELECTRICAL TECHNIQUES

Electrical Resistivity Methods.

A basic requirement of all electrical resistivity methods is that the borehole fluid (mud) be electrically conducting; this is necessary so as to give good electrical contact between the sonde (probe) being lowered and the formations. This requirement is exactly the opposite of the induction log (see later) requirement of a non-conducting fluid.

As with the ground based methods, we really determine an apparent resistivity, the weights given to various volumes of different resistivities depending upon the electrode geometry. Because of this, as a borehole sonde is is lowered through the formations (the equivalent of profiling at the surface whatever the electrode geometry), it does not "see" only the formation material which happens to lie between the two extreme electrodes, but also a large volume of material outside this distance; this is often referred to as the "adjacent bed effect".

The idealized radial distribution of resistivity around a typical borehole is shown in Figure 10.1; this is the equivalent of a depth distribution of resistivities (on a considerably smaller scale) in the ground based method (see Figure 2.9). Because of the complexity of the radial resistivity distribution a number of tools have been designed, each being particularly sensitive to part of the resistivity-depth spectrum.

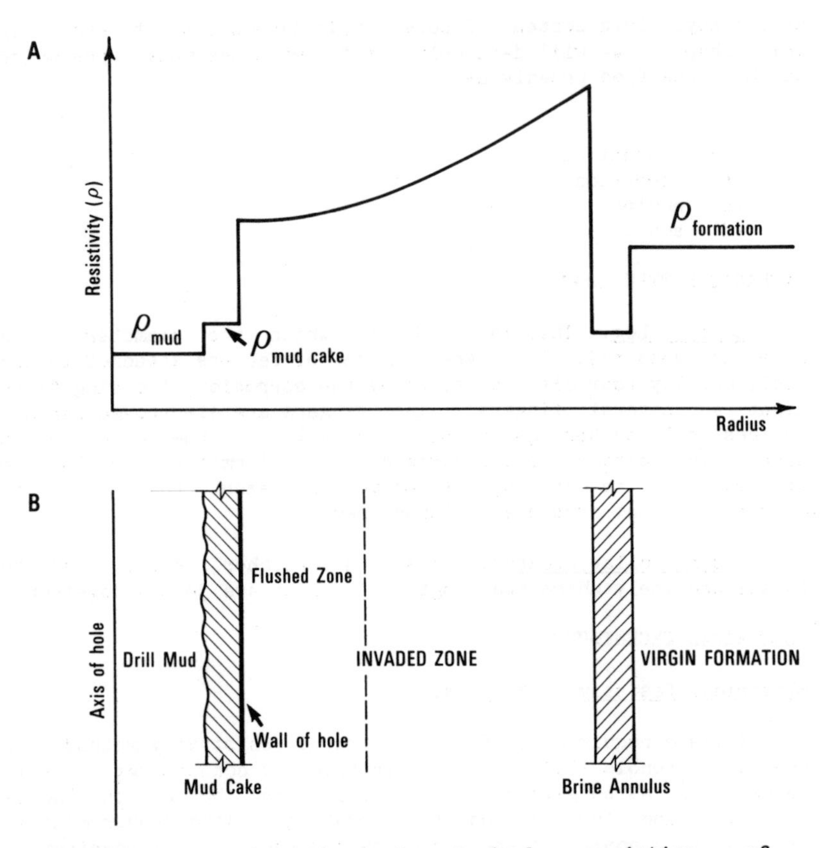

Figure 10.1. Schematic of typical variation of resistivity with radius from axis of a borehole in sedimentary formations. (A) The resistivity profile. (B) Origins of the resistivity variations. In the flushed zone the formation fluids are largely replaced by mud filtrate, the mud solids being "filtered" out to form the mud cake; the rest of the invaded zone contains progressively less mud filtrate. The brine annulus often occurs in oil bearing formations; since the hydrocarbons are more mobile than the saline component some separation occurs as they are pushed ahead by the filtrate.

Two electrode technique. This is frequently called a normal log and the electrode arrangement is shown in Figure 10.2(A); when the electrode separation is around 40 cm the arrangement is called a short normal log whereas when the electrode arrangement is about 150 cm it is called long normal log. The "normal log" has greater depth of penetration than the three electrode system, which will be described next, and gives symmetrical curves which are easier to interpret. Greater depth of penetration implies that the results are less affected by variations of the wall characteristics and also that the resolution is less than for shallow depth penetration instruments. If the bed thickness is greater than three times the electrode spacing, the "adjacent bed effect" is negligible and the true resistivities of the formations can be derived.

Three electrode technique. The arrangement is shown in Figure 10.2(B). When P_1 and P_2 have the same separation as C_1 and P_1 in Figure 10.2(A), the resolution is greater, and depth of penetration less, than for the two electrode technique.

For a long time these two techniques were obsolescent but recently there has been a revival of interest in their use because of questions raised about the true action of some of the more recent and more sophisticated devices.

Focussed current systems. Often called guarded log systems. Unfortunately there are so many trade names, some of them having been promoted to almost generic use, that the situation can be confusing. Where possible, alternative trade names for similar equipments will be given.

In a focussed current system the idea is to arrange guard electrodes in such a way that, by controlling with a feed-back

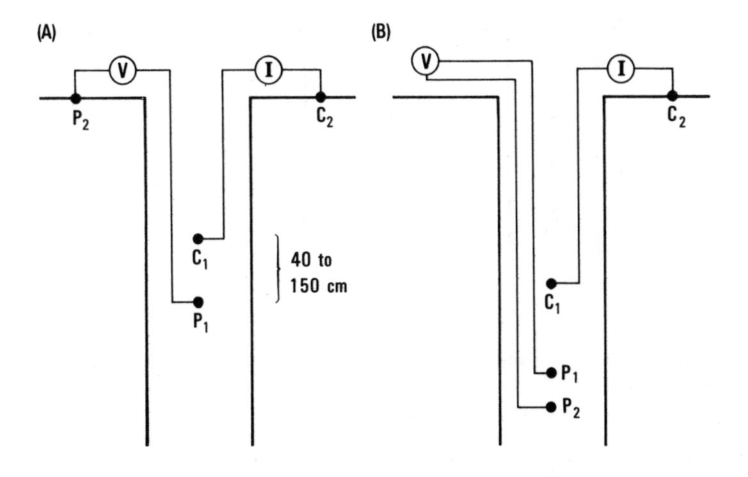

Figure 10.2. Schematic of resistivity logging method. (A) Two-electrode system. (B) Three electrode system.

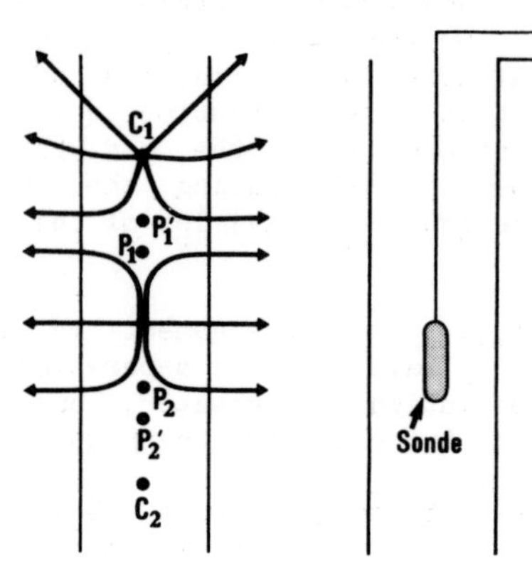

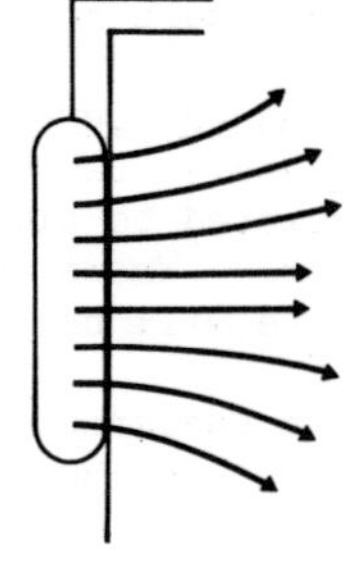

Figure 10.3. Schematic of focussed current systems. (A) Separation of C_1 and C_2 usually one or two metres and sonde is freely lowered; principal contribution to apparent resistivity comes from virgin formation radius range. (B) Spacing of electrodes on the sonde is of the order of a few centimetres and the sonde is pressed against the wall of the hole; principal contribution to ρ_a comes from flushed zone. (C) Spacing of electrodes is 2 or 3 cm and sonde is pressed against the wall of the hole; principal contribution to ρ_a comes from mud cake.

mechanism the current to various source electrodes, the current flow into the ground from the electrodes of interest is as nearly as possible at right angles to the logging sonde. For example, in the "Laterolog 7", a Schlumberger trade name, Figure 10.3(A), C_1 and C_2 are the "bucking" electrodes. Sufficient current is fed into C_1 and C_2 by using sensing electrodes $P_1P_1^1$ and $P_2P_2^1$ to indicate how much current is needed to ensure that the current flow from C_o, the central current electrode, is horizontal; clearly if this condition occurs the potential difference between $P_1P_1^1$ and between $P_2P_2^1$ will be zero. Other manufacturers use similar names such as "Guardlog", "Focussed Log" etc.

To estimate the resistivity of the flushed zone a button like sonde, about 10 cm in diameter, with concentric electrodes (up to seven of them) is pressed against the wall of the hole. As before the current flow from two current electrodes is controlled so that the current flow lines at the centre of the button initially form a cylinder but end up as a flaring, divergent cone, Figure 10.3(B); because of this the log is sometimes known as a trumpet log. Because of the small electrode separation the depth of resolution is therefore small and the principal use of the instrument is to estimate the resistivity of the "flushed zone".

Another type of micro resistivity sonde uses three electrodes spaced, in a line, about 2 or 3 cm apart and pressed against the wall of the hole, as shown in Figure 10.3(C).

Because ΔV is measured over a small depth interval, the value of ΔV is controlled by shallow phenomena such as mud cake, and this component of the sonde essentially "looks" horizontally (inverse resistivity), whereas the voltage V is essentially "looking" vertically (normal resistivity). The difference between V and ΔV (mud cake has a lower resistivity than the formation) is called the separation and is an indication of the amount of mud cake present. A caliper log is usually run at the same time. Numerous other types of sonde can be used with electrode arrangements designed to maximize the information from a particular radius range around the hole.

Continuous Dip Meter.

This is used to give (a) structural information and (b) sedimentary petrofabric information.

The dip meter consists of two main parts (1) three sets of electrodes of the microfocusing type set at 120° angular separation around the axis of the dip meter (2) a set of electro-mechanical devices to give the azimuth of one electrode with respect to magnetic north (from which we can get true north), the inclination of the borehole and the azimuth (or drift) of the inclination of the borehole. Frequently the instrument will contain a caliper device as well so that seven curves are obtained simultaneously.

Structural use (formation structure). The first section of the instrument is used to detect electrical resistivity differences as the instrument passes a boundary. If all three sets of electrodes show changes at the same time the structure is obviously normal to the axis of the borehole; if the resistivity changes at different times, Figure 10.4, then the structure has dip relative to the borehole axis. There are, of course, interpretational problems when crossbedding occurs but these can usually be solved.

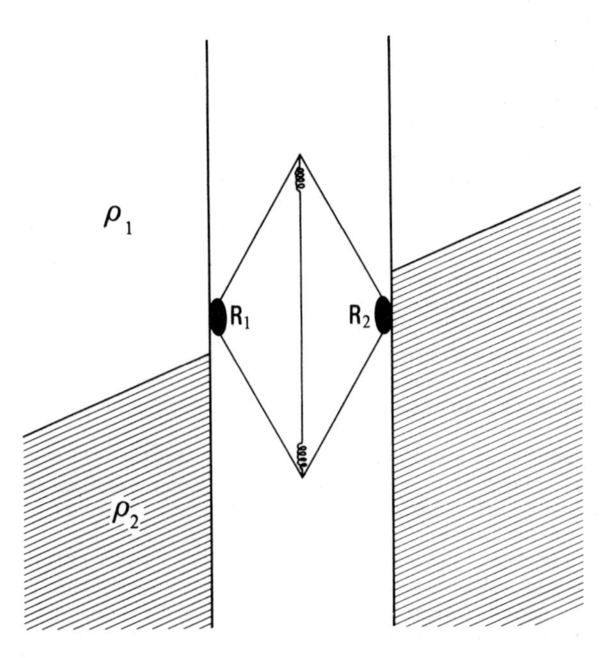

Figure 10.4. Principle of the continuous dip meter.

The second part of the instrument merely orients everything in space. The main reason for using this instrument is that valuable information can be obtained regarding structural features from just one hole. More reliable data could be obtained from a minimum of 3 holes but holes are expensive to drill just for this purpose.

Petrofabric use. This relies on the anisotropy of resistivity due to orientation of grains within the sedimentary formation.

Induction Log.

This is a profile of the electrical conductivity or resistivity obtained by a method utilizing electromagnetic waves. A high frequency is used (a few thousand Hz) with the receiver and transmitter being separated by up to a few metres and placed on a common axis. Some sondes may have more than one transmitter and receiver coil so as to produce a more focussed electromagnetic field.

A dual induction log is produced by a sonde which is operated at two different frequencies to obtain two different depths of penetration, with the results being indicated as ILd (induction log deep) and ILm (induction log medium). In all cases the sonde is lowered freely.

The current flows cylindrically whereas in the resistivity log it flows radially; cylindrical annulii are in parallel and therefore relatively low conductivity mud is required, otherwise all induced current would flow in the mud and simply give a result dominated by mud conductivity variations.

Self Potential Log.

A log can be made of the potential difference between an electrode in the hole and a reference electrode at the surface. Sometimes the potential difference between a pair of borehole electrodes, a few metres apart, is measured in which case we obtain a log of the oxidation potential.

The fundamental difference between borehole logging and ground surveys of natural potentials is that whereas in the latter case we are looking for mineralization potentials (see Chapter 3), the borehole device measures electrochemical potentials which are present at the interface between permeable beds, with the most striking contrasts occurring when porous sandstones are adjacent to shale formations, Figure 10.5. These potentials, regarded as background potentials or noise in a ground survey, are due to electrolytes of different concentration in contact, electrokinetic (or streaming) effects, etc. and range up to about 20 millivolts.

The method requires a salinity contrast, i.e different activities or ionic mobility, between borehole mud and formation waters. Therefore the logs cannot be run in a dry hole nor in an oil based-mud filled hole which has a low resistivity; on the other hand if the mud is too conductive it acts as a short circuit to the charges which give rise to the self potentials so that the mud filtrate must be carefully chosen. The activity is related to concentration via the activity coefficient.

Because of the low voltages involved, care has to be taken to

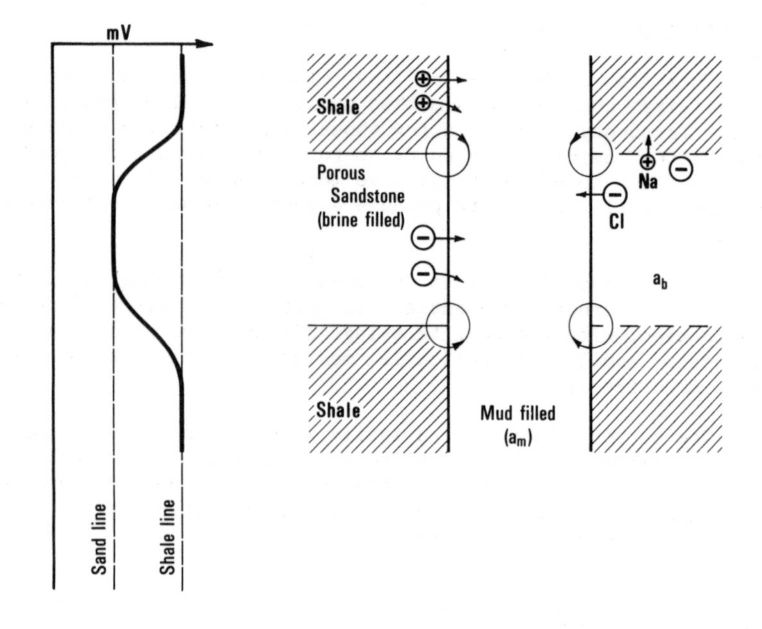

Figure 10.5. Principles of generation and variations in S.P.

reduce the effect of potentials arising from sources other than the formations. The principal problem here is the polarization of the logging sonde. This is usually made of an inert conductor such as lead and may become coated with a thin film of insulator so that the sonde can only handle a certain current density which, if exceeded causes the electrode to become polarized; that is, charges build up on the surface because they cannot be conducted away fast enough. In this situation the potential does not remain stable and the results are erratic.

The results of an SP survey are affected by temperature because viscosity and resistivity are temperature dependent.

Not only is the SP method a useful correlation tool but it can be used to give considerable structural - time related information, such as whether the formations are transgressive or regressive.

Induced Polarization Log.

This is based on the same principles as the ground type survey and is one of the very few methods in fairly regular use in the mineral exploration industry. However, it might also be useful in oil and gas wells because of the IP effects which often arise when clay is present.

NUCLEAR BASED LOGS

Nearly all types of nuclear based logs can be run in cased holes giving them a distinct advantage under some circumstances.

Gamma Ray Logs.

This is a record of the total count of natural radioactivity as the probe descends the hole. The detector is usually a NaI[Tℓ] crystal with electronics modified for borehole use. The log reflects mainly the shale content of the formations since the minerals containing the radioactive isotopes tend to concentrate in clays and shales; because of this the method is often used as a substitute for the SP log when a run has to be made in a cased borehole. The gamma ray log is used mainly as a correlation tool.

Gamma Ray Spectrometer Log.

This is similar to the total count gamma ray log but the equipment has at least three separate channels to record the Uranium, Thorium and Potassium contributions. More recently, because of the considerable interest in Uranium deposits, there has been a great deal of interest in a multi-channel spectrometer (MCS) logging device to try and obtain quantitative information on Uranium content and the MCS is likely to become an important mineral industry tool.

Special techniques can also used to locate the top of the cement, which has been laced with radioactive material, behind the casing.

Gamma-Gamma Log or Density Log.

This consists of a gamma ray source (e.g. Cesium-137) and a detector so shielded that it records only back scattered gamma rays - i.e. the Compton scatter. Because the back scattered radiation depends on electron density, see Figure 9.3, the count rate is roughly proportional to bulk density.

Source and detector are a few centimeters apart and pressed against the borehole wall. There may be a couple of detectors, the one with the largest separation from the source giving information weighted to larger radii. The short spacing detectors respond more to mud cake and small borehole irregularities and these readings can be used to correct the main detector readings for wall effects.

Nuclear Cement Log.

This is basically the same as the Density log but the source - detector separation is arranged so that the tool responds mainly to material in the annulus between the casing and the wall of the hole; it is used for distinguishing between cement and fluid behind the casing.

Neutron Logs.

There are various types of neutron logs with two principal uses (a) to measure hydrogen density, from which porosity can be estimated; (b) to measure chlorine density, from which salinity can be estimated.

Basis of method. Fast neutrons are emitted by a source in the sonde, this source being a mixture of radium and beryllium, plutonium and beryllium or even a special Van der Graaf generator. The neutrons then undergo three types of collision.

(i) Elastic scattering, where the neutron imparts some of its energy to the nucleus it has hit.
(ii) Inelastic scattering, where the neutron imparts enough of its energy to leave the nucleus it has hit in an excited state, the nucleus then returning to the ground state by emitting a characteristic gamma ray.
(iii) Neutron capture, where the neutron is captured to form a compound nucleus which then either (a) disintegrates, (b) emits a γ-ray to go to the ground state of a new elemental nucleus (c) emits a β-particle (with or without gamma rays) to form yet another ground state nucleus.

From nuclear theory it can be shown that on average a neutron loses more than half its energy on collision with a proton, which has a similar mass, but only a very small fraction for nuclei of heavy atoms. The mechanical analogue of this is best illustrated by thinking of what happens when two balls collide. If the projectile is small compared with the stationary ball it will simply bounce off with little loss of velocity and move the target only a small distance; on the other hand if the target and projectile are the same diameter, such as with billiard balls, the projectile loses much of its energy to the target and therefore bounces off the target with a much reduced velocity. Once the neutron has been slowed to thermal energies (approximately 0.025 electron volts), i.e. to velocities at which neutrons move thermally, it is readily captured. The ability of a particular nucleus to capture neutrons (and other particles) is related to its capture cross section.

A common element in boreholes is hydrogen which has an atomic mass of 1; molecules containing hydrogen therefore scatter and slow down the neutrons very efficiently. Once this has happened, chlorine captures neutrons more readily since its capture cross section for neutrons is 100 times that of hydrogen, while those for carbon and oxygen are 100 and 1000 times smaller than for hydrogen. At the time of capture the chlorine gives off a characteristic gamma ray of nearly 7 MeV.

The neutron log can be adjusted to record

(a) all capture gamma rays (n-γ) - when the log is essentially recording hydrogen density;
(b) the characteristic gamma ray of thermal neutron capture by chlorine (n-n); or
(c) epithermal neutrons, those just above the thermal velocities.

The essential difference between a chlorine and a hydrogen log lies in the type of gamma rays they are set to record. Since the systems are so similar, obviously one will effect the other and both are usually run together to put in appropriate corrections to the other.

These logging techniques can be used in cased boreholes and frequently run by the trade name such as Chlorinilog, Saltilog, Salinity log, etc.

Hydrogen density log. Referring to figure 10.6, neutrons are emitted by source S and travel into the surrounding formations. If the hydrogen content near the source is high the neutron is soon stopped and the resultant gamma ray has to travel a fair distance to the detector; therefore it is readily absorbed after constant scattering so that with high hydrogen concentration near S the total count rate at detector D is low.

Conversely, when the hydrogen concentration is low, the neutrons travel a fair distance before being captured; the resultant gamma rays have only a short distance to travel and reach the detectors before being stopped. Therefore for low hydrogen content near S the total count rate at D is high.

Where formations are "clear", that is the hydrogen is not bound as in water of crystallization (found in shales and gypsum, for example), the measurement of hydrogen can be turned into a measurement of porosity by assuming the pores are filled with water or oil. A low hydrogen density indicates a low liquid filled porosity.

Neutron life time log. This differs from the previous technique in that it is designed to record the capture of thermal neutrons by chlorine by recording the count rate of 7 MeV gamma rays. Essentially, the technique therefore gives a log of the capture cross section of the formations, with particular sensitivity to chlorine. Since electrical resistivity is strongly dependent on the electrolytes present in the pores of the rock, of which salt is one, the chlorine log resembles a resistivity log and is sometimes run when the resistivity log cannot be run, for example, in a cased hole. Figure 10.7 indicates the pattern of results which can be obtained.

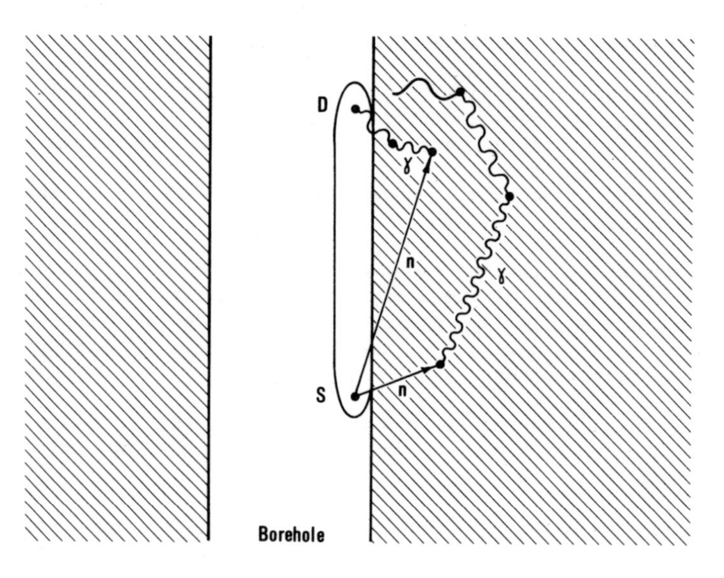

Figure 10.6. Hydrogen density log. When hydrogen density is high near the neutron source S, neutrons are slowed rapidly and travel only a short distance before being captured; on capture a gamma ray is emitted which then can be scattered many times without any energy reaching the detector. When hydrogen density is low near S the neutrons can travel further, thus giving the resultant capture gamma rays a higher probability of reaching D and giving a higher count rate.

Nuclear Magnetism Log (NML).

This basically gives a profile of hydrogen density. The instrument is based upon the same principles as the proton magnetometer (see Chapter 6) but in this case the fluid to be polarized is the fluid in the pore space close to the wall of the hole.

Essentially, a strong polarizing field is applied perpendicular to the axis of the hole to polarize the pore fluid nuclei, the sonde being held against the wall of the hole. When the polarizing field is switched off, the hydrogen nuclei start precessing as they try to align themselves with the earth's field. The precession frequency will identify them as hydrogen but the amplitude of the response and the decay time depends upon the number of nucleii per unit volume; either of these parameters then becomes a method of determining the "density" of hydrogen nucleii present. The result can be expressed as a free fluid

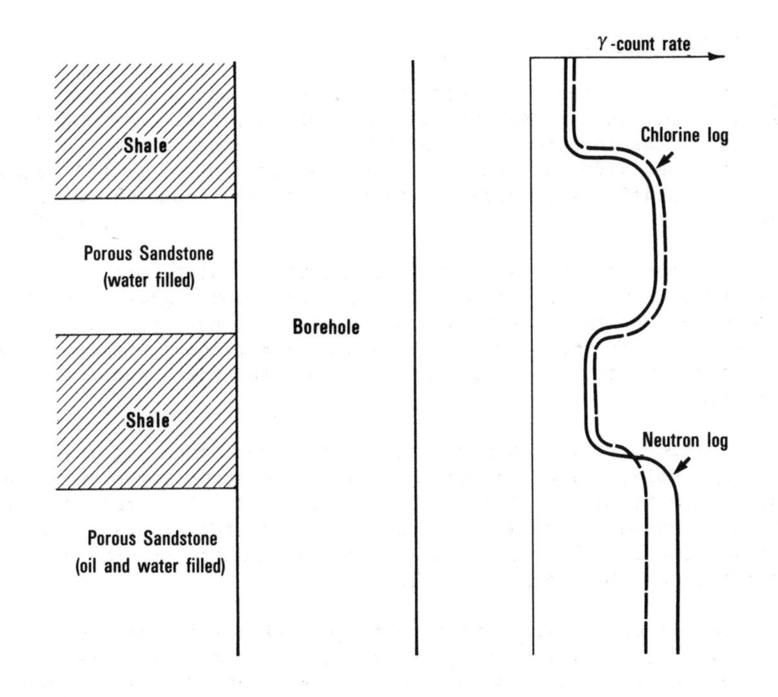

Figure 10.7. Idealised log of chlorine and hydrogen density, illustrating principles of interpretation.

index (FFI) since it has been found that the principal response comes from protons sufficiently free that they can be considered mobile and the fluids therefore potentially producible.

Protons that are bound in any significant way, such as in solids, or are subject to high surface forces, as in shales or near clay mineral surfaces which have a very high surface area, give a very attenuated response, i.e. they have a short decay time. Therefore hydrogen in water which is water of crystallization or held near the shales and clay does not respond as well as hydrogen in hydrocarbons existing as fluids in pores.

Typical decay times are 50 to 300 μs for water and greater than 600 μs for hydrocarbons. Therefore, by delaying the start of the measuring interval the interference due to the response of water is reduced to a minimum.

SONIC LOGS

There are two parameters of interest, the travel time and the attenuation of a seismic (acoustic) disturbance over a short length, up to a few meters, of hole.

Travel Time Methods.

The log is usually expressed in terms of the time taken to travel from the transmitter to the receiver, i.e. the output on the log is the inverse of velocity. Since the sonde is held against the side of the wall the results are unaffected by the "adjacent beds" effect because the path of the acoustic wave is direct.

The sonde can be a single transmitter with one or more receivers; the two or multi receiver sonde gives the time difference over small distances so that, since our principal interest is in the travel time in the formations themselves, the error due to travel time in the mud at both ends can be eliminated.

Sometimes it is not possible to press the sonde against the wall of hole, in which case a compensated sonic logger might be used. This is essentially the same as a sonic log but doubled with two transmitters which are pulsed alternately; averaging the readings for the time difference tends to cancel errors due to sonde tilt or changes in the hole radius.

The interval velocities can be integrated to give a velocity-depth profile which is useful for good interpretation of seismic profiling data obtained by surface methods (see Chapter 8).

Amplitude Methods.

Techniques based on this principle are used specifically for the cement bond log (CBL).

One of the big difficulties in bringing wells into production is to make sure that the casing has been properly cemented in place; we need to know how good is the bond connecting casing to formations.

If a single detector continuous velocity log is run in a cased (and cemented) hole the path of minimum time is usally via the steel casing. Travel time is therefore of little use in the cased hole; however, amplitudes can be used. Assume first that the velocity in steel is greater than the velocity in the formation.

If the hole is cased and not cemented (or is cemented with a poor bond between the casing and the cement) all the energy is channeled along the steel casing so the signal at the detector has a large amplitude.

If the casing is cemented with a good bond between the casing and cement but with a poor bond between the cement and the formations, much of the energy is transmitted via the cement and the first arrival at the receiver, via the steel, contains little energy and therefore has low amplitude. If the bond is good between the cement and both the casing and the formations, the energy is nearly all lost to the formations; therefore, although the first arrival is still via the steel it has so little energy it cannot be detected.

Sometimes (although very rarely) the transit times in the formations are shorter, not longer, than in the steel. In such circumstances the interpretation is different but can still be carried out, provided we record the transit time as well as the amplitude, as can be seen by looking at Figure 10.8; there is no signal from the steel but a formation signal will be received - earlier than if the energy had passed through the steel.

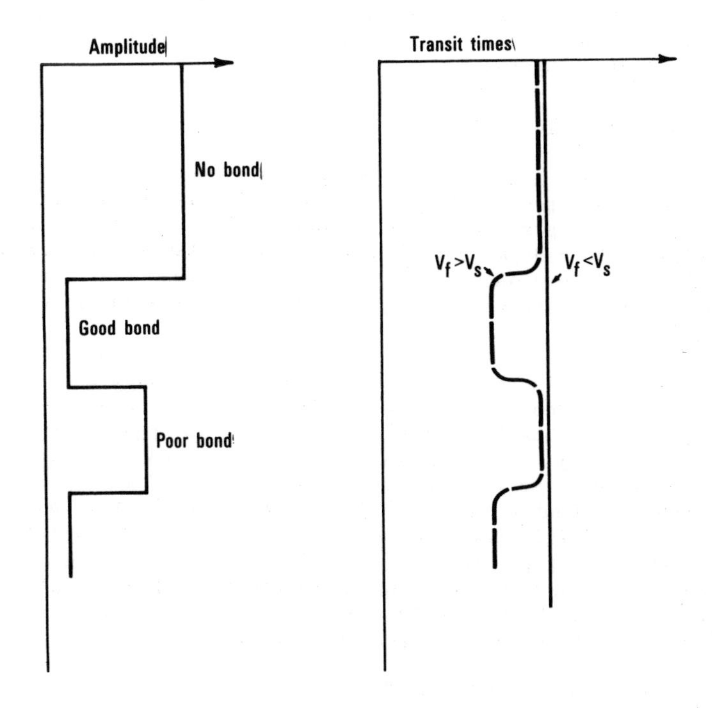

Figure 10.8. Idealised cement bond log (CBL) illustrating principles of interpretation.

OTHER TECHNIQUES OF INTEREST

Gravity Logging.

From time to time interest has been shown in using a borehole gravimeter to produce logs from which a density profile can be obtained. Because of the great sensitivity of the gravimeters the instrument has to be quite stationary when it is read and requires up to 30 minutes for each reading. With station spacings of 5 to 10 meters, density differences of 0.01 gm/cc can be resolved; this is equivalent to about a 1% change in average porosity.

Because the gravitational attraction at the meter is the sum of all masses within its sensitivity range, the results are not significantly affected by mud cake or near hole changes due to drilling in the formations, e.g. flushed zone - Figure 10.1, nor does casing affect the difference values since it has a constant influence at each station; rather, the technique may be used to detect bodies, such as salt domes, in close proximity to, but not penetrated by, the borehole.

At times, the differences between the gravimeter density log and the gamma ray density log can be used to give valuable structural information.

Thermal Logging Techniques.

Temperature logs are used mainly in a subsidiary way; since many physical properties, but especially electrical resistivity and viscosity, are temperature dependent, the data are used principally to reduce other logs to a common temperature reference level. For this reason no real attempt has been made to obtain accurate temperatures, errors of several degrees being not at all unusual.

However, with recent improvements in technology and borrowing some of the deconvolution techniques developed in seismology, it is possible to obtain temperatures to a few thousandths of a degree accuracy. This means that temperature gradients can be determined precisely, to within about 0.2 ^{o}C/km (mK/m) over distances of about a meter. For a borehole in thermal equilibrium this is essentially the same as logging thermal resistivity and since the results can be obtained equally well in cased and uncased holes, and the technique effectively samples large volumes of rock and is therefore insensitive to near hole variations, it may become another useful correlation tool.

SUGGESTIONS FOR FURTHER READING

Conaway, J.G. and Beck, A.E., 1977. Fine scale correlation between temperature gradient logs and lithology, Geophysics, 42, 1401-1410.

Conaway, J.G. and Killeen, P.G., 1978. Quantitative uranium determinations from gamma-ray logs by application of digital time series analysis, Geophysics, 43, 1204-1221.

Hearst, J.R. and McKague, H.L., 1976. Structure elucidation with borehole gravimetry, Geophysics, 41, 491-505.

Pickett, G.R., 1970. Borehole geophysics symposium, Geophysics, 35, 80-152.

Pirson, S.J., 1970. Geologic Well Log Analysis, Gulf Publishing, Houston.

Roy, A. and Dhar, R.L., 1971. Radius of investigation in D.C. resistivity well logging, Geophysics, 36, 754-762.

Wylie, M.R.J., 1963. The Fundamentals of Well Log Interpretation, Academic Press, New York.

Index

APPENDIX: Useful Aspects of the International System of Units (SI).

SI BASE UNITS

Quantity	Name	Symbol
length	meter	m
mass	kilogram	kg
time	second	s
electric current	ampere	A
thermodynamic temperature	kelvin	K
amount of subst	mole	mol
luminous intens	candela	cd

SI PREFIXES

Multiplying Power of 10	Prefix	Symbol	Multiplying Power of 10	Prefix	Symbol
18	exa	E	-18	atto	a
15	peta	P	-15	femto	f
12	tera	T	-12	pico	p
9	giga	G	-9	nano	n
6	mega	M	-6	micro	µ
3	kilo	k	-3	milli	m
2	hecto	h	-2	centi	c
1	deca	da	-1	deci	d

SOME USEFUL CONVERSIONS FROM COMMONLY USED UNITS TO SI

Where the derived SI unit has a special name, this is shown in the penultimate column with the approved symbol and unit in brackets. Where a letter is used both as a prefix and as a base unit, the prefix should not be followed by a space. For example, thermal conductivity (watts per meter kelvin) should be written as $Wm^{-1}K^{-1}$ or W/m K but not W/mK since this last form could be misread as watts per millikelvin; however, the last form may in fact be found in work where there is little possibility of misunderstanding by others in the field. The first form is probably the best one to use, although the second form is easier when manuscripts are prepared on a computer (such as this one).

Physical Quantity	c.g.s. Unit	Multiply by	to get SI Unit	In Chapter
<u>Electrical</u>				
Current	esu/s	3.335×10^{-10}	ampere (A)	2
	emu/s	10		
Energy (work,	calorie	4.184	joule (J - kg m^2 s^{-2})	
quantity of heat)	erg	10^{-7}		
Potential	erg/esu	299.8	volt (V - J A^{-1} s^{-1})	2,3,4
	erg/emu	10^{-8}		
Resistance	ohm	1	ohm (Ω - V A^{-1})	2
Conductance	mho	1	siemen (S - A V^{-1})	
Resistivity	ohm m	1	ohm m (V A^{-1} s)	2,4,7
Conductivity	mho/m	1	S m^{-1}	2,4,7
Inductance	emu	10^{-9}	henry (H - V A^{-1} s)	7
<u>Magnetic</u>				
Susceptibility	dimensionless	4π	dimensionless	6
Permeability	dimensionless	$4\pi \times 10^{7}$	H m^{-1}	6
Flux	maxwell(Mx)	10^{-8}	weber (Wb - V s)	6
Flux density	gauss (G)	10^{-4}	tesla (T - V s m^{-2})	6
	gamma (γ)	1	nT.	6
Field strength	oested (Oe)	$1000/4\pi$	A m^{-1}	6
<u>Other</u>				
Velocity	m/s	1	m s^{-1}	8
Acceleration	gal (cm/s^2)	1	Gal (cm s^{-2})	5
Force	dyne (gm cm/s^2)	10^{-5}	newton (N - m kg s^{-2}.)	5,6
Pressure	bar (dyne/cm^2)	10^{5}	pascal (Pa - dN m^{-2})	
	torr	101 325/760	Pa (Torr)	
Power	erg/s	10	watt (W - J s^{-1})	
Activity (radioactivity)	curie (Ci)	37×10^{-9}	bequerel (Bq - s^{-1})	9
Frequency	cycles/s	1	hertz (Hz - s^{-1})	4,7,8
Electron volt	eV	16×10^{-20}	joule (J)	9,10
Thermal conductivity	cal/s cm °C	418.4	W $m^{-1}K^{-1}$	
Temperature	°C	subtract 273.15	kelvin (K)	